Metrics and Methods for Security Risk Management

Metrics and Methods for
Security Risk Management

Carl S. Young

AMSTERDAM • BOSTON • HEIDELBERG • LONDON
NEW YORK • OXFORD • PARIS • SAN DIEGO
SAN FRANCISCO • SINGAPORE • SYDNEY • TOKYO
Syngress is an imprint of Elsevier

ELSEVIER

SYNGRESS.

Acquiring Editor: Pam Chester
Development Editor: Greg Chalson
Project Manager: Paul Gottehrer
Designer: Alisa Andreola

Syngress is an imprint of Elsevier
30 Corporate Drive, Suite 400, Burlington, MA 01803, USA

Notices
Knowledge and best practice in this field are constantly changing. As new research and experience broaden our understanding, changes in research methods or professional practices, may become necessary. Practitioners and researchers must always rely on their own experience and knowledge in evaluating and using any information or methods described herein. In using such information or methods they should be mindful of their own safety and the safety of others, including parties for whom they have a professional responsibility.

To the fullest extent of the law, neither the Publisher nor the authors, contributors, or editors, assume any liability for any injury and/or damage to persons or property as a matter of products liability, negligence or otherwise, or from any use or operation of any methods, products, instructions, or ideas contained in the material herein.

Library of Congress Cataloging-in-Publication Data
Application submitted

British Library Cataloguing-in-Publication Data
A catalogue record for this book is available from the British Library.

ISBN: 978-1-85617-978-2

For information on all Syngress publications visit our website at www.syngress.com

To my parents, Dr. Irving Young and Dr. Geraldine Young, from whom I learned what is and is not worth risking.

Table of Contents

About the Author

Carl S. Young is a recognized international authority on risk management and security technology. He held a senior position in the U. S. government and has been credited with significantly enhancing the U. S. capability in special technical methods. From 1999 to 2002 he was a consultant to the JASON Defense Advisory Group and served on a blue ribbon panel on technology appointed by the Director of Central Intelligence.

He is currently a Vice President and senior risk strategist for a major international corporation as well as an adjunct professor in Protection Management at the John Jay College of Criminal Justice (City University of New York). In 1997 he was awarded the President's Foreign Intelligence Advisory Board (PFIAB) James R. Killian Award by the White House for individual contributions to national security.

Mr. Young holds undergraduate and graduate degrees in mathematics and physics respectively from the Massachusetts Institute of Technology (MIT).

Foreword

The 9/11 attacks on the Twin Towers, the Pentagon, and (nearly) the White House are a landmark for those involved in security management in the government or private sector. Even before 9/11 the most forward-looking organisations, faced with sophisticated criminal and terrorist threats, had begun fundamentally re-engineering their approaches to security. The old, quasi-military, approach of 'guns, guards, and gates' was exposed as expensive, inflexible, and ineffective. It had failed and served only to give a false sense of security. From the embers of 9/11 a new risk-based approach to security gained traction. Security ceased to be an art and became a science. Guesswork was replaced with careful analysis based on scientific methodologies.

I was fortunate to be part of that process in government and in the global financial sector. In both I worked closely with, and relied heavily upon, Carl Young. In the post-9/11 world we had to find cost effective, practical, risk-based, resilient solutions to immensely challenging issues. Carl Young was, and is, central to that work. He combines academic brilliance with practical, hands-on experience of delivering security solutions. This book is a synthesis of that work. It provides a sound basis on which to make a wide range of security judgments with confidence. Moreover it is highly readable. I recommend it to the reader.

James A. King, CBE
Director of Intelligence, Physical and Personal Security,
Group Security and Fraud, Lloyds Banking Group

Preface

Believe it or not, some of my earliest moments on the planet were spent in the company of my parents while they toiled away on human cadavers. I doubt this was a traditional form of family entertainment, especially in the 1950s. But as both newly-minted parents and clinical pathologists they juggled their careers with domestic obligations as best they could. It seems that decent baby sitters have always been in short supply. There is no telling what effect this experience had on their eldest child's development or whether it influenced future career decisions, but it probably does help to explain personality traits that are probably best explored elsewhere.

I consider myself fortunate to be working in security risk management, which has clearly been at the forefront of public awareness since September 11, 2001. Some might find it ironic that the events of that day caused a huge uptick in an interest in security almost overnight. The irony is twofold: terrorism has been around for a long time (recall Guy Fawkes in 1605) and there is now a focus on security in ways that have nothing to do with terrorism. One possible explanation is that this horrific event exposed a powerful nation's vulnerability and raised the specter of much broader security concerns.

In my view, the consequences of this renewed interest are mixed. On the positive side, corporate security is no longer viewed as a necessary evil and left to be managed in relative obscurity by non-professionals. Progressive firms now view security as part of the company's business strategy. Savvy executives even market security as a means of distinguishing their company from the competition.

The downside has been the inevitable increase in "security theater", a term purportedly coined by the cryptography expert Bruce Schneier. These are measures that give the appearance of providing security but are ineffective when exposed to rigorous analyses. The field of security tends to be dominated by action-oriented types who sometimes invoke a "ready, shoot, aim" approach to problem solving. That is okay if the goal is just to get something done quickly. Unfortunately without a coherent and reasoned approach to risk it is not clear that "something" is always effective.

Security problems in the commercial world have changed in part because the office environment itself has evolved. These changes are due principally to the proliferation of the Internet as a communication tool in conjunction with ubiquitous software applications that facilitate the creation, transmission, and storage of information. These technology advances represent security challenges precisely because they are integrated into the fabric of companies at every level and make communication incredibly convenient from almost anywhere in the world.

Risk mitigation is of great importance to modern corporations. However, a truly useful mitigation strategy is one that is derived from a big-picture perspective and realistic approach. An aggressive security posture might be effective but can't be at the expense of business performance. Aside from hurting the bottom line, such a strategy could result in a one-way ticket to unemployment for the well-intentioned security director.

In today's world, private companies are often viewed as representatives, if not ambassadors, of the countries in which they are incorporated and/or physically located. So not only are companies sometimes targeted by competitors in order to steal their information, they are also the focus of political or religious groups who understand their economic and symbolic importance.

At the same time, budgets are decreasing while security departments are dealing with threats that demand greater vigilance and resources. In the wake of the 2008 global economic meltdown, corporate executives are asking more difficult questions about return on investment. But the effectiveness of the defensive measures used in security is difficult to quantify in the same way as profit and loss. That is part of what this book is all about. The need for rigor in security is greater today than ever and not only to address more complex threats, but also to employ cost-effective methods that are explicitly proportionate to risk.

This book attempts to bridge the worlds of two distinct audiences. One group consists of career security professionals who have wisdom born of experience in assessing risk but often possess no technical background. In the other camp are the scientists and engineers who work on technical problems related to security but have little or no background and therefore lack the context for these specialized problems. The former group often knows a lot about security but has little technical knowledge. The latter group has familiarity with mathematics and/or scientific principles but may not know how these apply to security risk.

Many individuals who work in security function as both theorist and practitioner. This is a difficult challenge in a field where the theoretical underpinnings have not been formally recognized or at the very least have not been centrally codified. It is precisely this divide between theory and practice that must be solidified for security professionals to continue to grow and if the subject is to be universally accepted as a legitimate academic discipline.

It is important to recognize that security problems must be viewed in terms of risk in order to be relevant to the corporate world. Although significant insights will be gained from the study of well-established physical principles, the utility of these principles derives from knowing how they affect risk, and moreover, how they can be used to develop effective and proportionate mitigation.

To that end, this book endeavors to provide the reader with the following: the fundamentals of security risk and its individual components, an analytic approach to risk assessments and mitigation, and quantitative methods to assess the individual components of risk and thereby develop effective risk mitigation strategies. In so doing, I hope it will provide security professionals, engineers, scientists, and technologists with both an interesting and useful reference.

This book is divided into two distinct parts. Part 1 is entitled "The Structure of Security Risk" and comprises Chapters 1 to 3. Part 2, "Measuring and Mitigating Security Risk", consists of Chapters 4 to 6.

Part 1 is meant to be a detailed exposition of security risk and I believe it is a unique treatment of the subject. It discusses the individual components of risk in detail as well as some important physical models relevant to assessing those components. These will be crucial to the development of the risk metrics discussed in Part 2. In addition, risk assessment and mitigation processes are delineated and can assist in establishing a risk-based security management program.

Specifically, the fundamentals of risk management are discussed in Chapter 1. In particular it introduces a key expression that I somewhat dramatically refer to as "The Fundamental Expression of Risk." This important statement expresses the defining attributes of risk and is fundamental to any problem in security. In particular, the likelihood and vulnerability components are discussed in detail and are the focus of much of this book. This chapter also discusses the role of important tools such as security standards and risk models.

Chapter 2 introduces key security-related concepts that are used to measure risk and thereby establish security metrics later in the book. It discusses the notion of scale or how physical quantities that affect the vulnerability component of risk change as a function of scenario-dependent parameters like distance and time. Recurring physical models are highlighted that directly relate to the assessment and mitigation of the vulnerability component of risk and are discussed in detail in Part 2.

Chapter 3 may arguably be the most appealing and/or useful to security professionals. It describes the risk assessment and risk mitigation processes in detail. These provide the context for the technical methods discussed in Part 2. This chapter also specifies how the risk mitigation process provides a natural segue to the development of risk-based security standards, assessments, metrics, and security program frameworks.

At this point I must give even the most intrepid reader fair warning: Part 2 represents a more quantitative treatment of security risk management than security professionals may be accustomed. However, Part 2 provides the machinery that is necessary to rigorously assess security risk and that has been mostly absent from traditional books on security. The goals are two-fold: to show the engineer or scientist how well-established scientific principles apply to security risk problems and to introduce the security professional to key technical/scientific concepts that are important to assessing security risk. Wherever possible, real-world examples are provided and sample calculations are performed.

Chapter 4 provides the concepts and techniques necessary to assessing the likelihood component of risk. These include useful probability distributions and a discussion of the important distinction between the *likelihood* and *potential* for further incidents. The goal is to provide the reader with an appreciation for some of the probabilistic tools that are relevant to security risk and to show how and when they apply.

Chapter 5 details the physical models, principles, and quantitative methods necessary to assess the vulnerability component of risk. The recurring physical models and mathematical functions introduced in Chapter 2 are applied to a variety of security risk scenarios. In addition, numerous calculations that yield quantitative risk metrics are spelled out. Although the mathematics and physical concepts are straightforward and well-understood by scientists and engineers, their applicability to problems related to security risk has not been articulated nor fully appreciated within the security community.

Finally, Chapter 6 applies the methods of Chapters 4 and 5 as well as the results of published and unpublished studies to address a variety of security threats. The intent is to provide more in-depth analyses of security risk mitigation as well as give quantitative if ballpark estimates of the efficacy of these methods. At the very least, such analyses can be used as a reference for discussions with consultants pursuant to obtaining a reality check on proposed solutions.

Acknowledgments

With respect to acknowledgments, I must first mention my parents, Irving and Geraldine Young, who along with providing years of unconditional emotional support also paid my college tuition. I will be forever grateful for the love and understanding they lavished upon their undeserving son.

My siblings, Diane Uniman and Nancy Young and their respective spouses Howard Uniman and Jeff Melin, are next in line in terms of acknowledgment. Although their contributions to this book may not be obvious, they are nevertheless profound and traceable to my earliest days. I am extremely fortunate to have such a supportive family.

My sometime collaborator, David B. Chang, is a brilliant physicist who is able to take a germ of an idea and develop a much deeper understanding of risk. In particular, I must acknowledge Dave's detailed editing as well as his substantive contributions to the corpus of material. If this book is successful it is primarily due to his significant contributions.

My interest in security actually began sometime during challenging if undisciplined days in Senior House at MIT. I to have remained close to two friends from that time, David and Lisa Maass. Their encouragement and support over many years and especially since my arrival in New York has directly contributed to successes in my career as well as in the evolution of this book.

My many years in Washington were instrumental in developing my thinking about security. Specific individuals profoundly contributed to my security education and in the process became close friends. Most notably these include Marc Cohen, Steve Dillard, Tim Houghton, Dave Murphy, and Martial Robichaud.

David Tinnion, a former mentee, challenged me and others to think precisely and broadly about security risk. I am proud to say that in his case the student has surpassed the teacher. Another colleague, Caroline Colasacco, is a brilliant analyst upon whom I have relied on for her insight and native problem-solving ability. Hopefully these two individuals will be directors of security but will no doubt be transformative figures if given the right opportunities.

I must thank Kristine Ciaburri (aka "Ciaburrus"), who provided expert advice on the subtleties of Microsoft Word as well as a range of administrative issues. She is one of the most proficient administrative assistants in the business despite the torrent of rubber bands I continue to fire her way.

Bob Grubert, a recognized physical security expert, has been an important sounding board for many of the ideas that appear in this book. Thankfully, he has never shied away from giving advice in ways that are all but impossible to ignore.

Peter Rocheleau, my friend for nearly four decades, generously contributed his time and expertise in reviewing this manuscript. He also provided much-needed insights into the world of publishing.

My cousin, Kate Smith (no relation to the singer), has been a staunch supporter and proponent of this book. Our dinners in Brooklyn not only cheer me up but always give me plenty of food for thought.

Ruth Steinberg, Manhattan's premier physician and culture maven, has sustained my writing efforts with her cooking and robust conversations during our "Sunday Séances." She was also my crossing guard in elementary school and could arguably be considered my first security mentor.

The staff at Elsevier have been a delight to work with. In particular, I must acknowledge the efforts of acquisitions editor Pam Chester and production manager Paul Gottehrer. They each deserve the publishing equivalent of the purple heart for their extreme patience and professionalism.

Certain individuals in academia and industry greatly helped to shape my thinking about security over many years. In particular, Sid and Harriet Drell, Dick and Lois Garwin, Paul and Vida Horowitz, and Mal and Paula Ruderman have all been important mentors and remain some of my closest friends. I owe considerable thanks to Jim King for giving me the freedom to pursue non-traditional paths that led to a number of the concepts that appear in this book. He is a true visionary in the art and science of security risk management as well as a great friend.

I conclude with a quote from Mark Twain, who captured the essence of security risk management in one cautionary phrase: "Put all your eggs in one basket...and better watch that basket!"

The Structure of Security Risk

Security threats and risk

Threats are illogical. — Sarek.

"Journey to Babel," *Star Trek*, **Stardate 3842.3**

1.1 INTRODUCTION TO SECURITY RISK OR TALES OF THE PSYCHOTIC SQUIRREL AND THE SOCIABLE SHARK

Ask a hundred people to state the difference between threat and risk and you will likely get a very diverse set of answers. I often ask this question when interviewing candidates for a security-related job. Even those who assess risk for a living are often stumped when asked for a working definition of these two terms. To complicate matters, colloquialisms abound to include "managing risk," "risk relevance," "concentration of risk," "risk free," and "risk averse." Each implies something tangible if not downright quantifiable.

Many of us rely on intuition biased by cultural norms and wishful thinking to make important decisions that one might broadly characterize as "risky." These decisions include buying and selling property, taking a job versus attending school, and even (or especially) getting married. Is intuition alone sufficient to reach appropriate decisions in these matters? More generally, how well equipped are we to assess risk and how successful are we at doing so?

Although engaged couples probably do not want to think of marriage as a problem in risk management, one might get a completely different perspective from the legions of divorced couples. Let's examine what I unromantically choose to call marital risk. More precisely, we wish to examine the likelihood that a marriage contract between two individuals (any gender) will be terminated before death does them part.

Unfortunately I am not certain there is a reliably predictive model for marital risk factors. I instinctively feel such deliberations are better left to the numerous books on interpersonal relationships found in the self-help sections of bookstores. Most couples probably assume the standard marital problems apply to everyone else as they blissfully stroll down the aisle and enter the ranks of a very robust statistical model.

However, it may be relevant to point out that according to generally accepted but admittedly anecdotal reports, between 40 and 50% of all marriages in the United States end in divorce. Despite the universal optimism on the part of newly married couples, divorce lawyers continue to make a nice living. Does marriage carry a special form of risk or is the decision to marry so burdened by emotion that people just ignore the risk factors? Is each case so unique that historical data are irrelevant? The reliability of predictions on the likelihood of future incidents based on historical precedent is an especially critical issue and is often encountered in security risk scenarios.

There seems to be little question about the right way to mitigate the financial risks associated with marriage: the arguments for a prenuptial agreement seem downright compelling. This is especially true if one or both parties have any appreciable assets. Imagine contemplating another type of business arrangement where there was a 40 to 50% historical likelihood of failure. Would you hesitate in obtaining some form of insurance, especially a policy with relatively low premiums that clearly offers an effective hedge against potential losses?

Hopeless romantics might argue that the presence of the agreement enhances risk. Possibly so, but a low-cost agreement that limits the exposure to financial loss certainly sounds like a good idea. This notion probably resonates more with those who have been divorced and have the scars to prove it. Why are these agreements not standard practice? Why is there such resistance to the idea? One guess is that emotion has a vitiating effect on some risk decisions. Sadly, I believe that most marriages utilize a common and flawed risk mitigation strategy of assuming a low likelihood of threat occurrence (i.e., divorce) at the expense of threat vulnerability. This very issue will be discussed in more depth when the psychotic squirrel and the sociable shark are introduced.

Let's take a step or two back in time and speculate on how our ancestors may have dealt with decisions that were critical to their survival. Our prehistoric forebears probably had to do some serious risk assessing when facing the threat of a hungry saber-toothed tiger. It might be surmised that on average our distant relatives made the better risk decisions since we are still around and the poor feline has been relegated to a fossilized relic.

In fairness, other forces probably contributed to the animal's demise, but it is clear that humans have been the more resilient of the two species.

Even if such judgments served humans well thousands of years ago, it is not clear that the "fight or flight" instinct is particularly relevant to humans nowadays. In the old days, the consequences of a bad choice were clear-cut, and over time, the forces of evolution were unforgiving. Less adaptable species succumbed to extinction. Although possessing inferior physical prowess when compared to the tiger, humans shifted the odds in their favor through superior intellect. They developed weapons and techniques that were incorporated into increasingly successful hunting strategies. In the process they might have developed an intuition about the risk of confrontations in general. Some of their less cerebral adversaries continued to rely on instinct for survival, and instinct alone for some species proved to be a losing strategy in the face of unassailable forces.

Humans thrived as other species became extinct through predation, changing climates, natural disasters, etc. This might support the notion that humans continue to be relatively adept at decisions on risks that impact physical survival but are not so adept at other risk decisions. It would be interesting to contemplate the effect on our approach to marital risk if lousy marriages resulted in a 40 to 50% death rate rather than divorce. I suspect those who modified their behavior would be the surviving members of the species and evolve into individuals highly attuned to marital risk.

Consider a common physical threat that we face each day; namely being hit by a car. How many times have you crossed the street in the midst of traffic or even flagrantly ignored the "Don't Walk" signal and miraculously lived to tell about it? Somehow you successfully determined whether you were likely to get flattened and acted appropriately. Maybe it is just not that difficult a problem, although judging from the prevalence of road kill it seems that less enlightened species often get it wrong.

Perhaps we are attuned to the relevant risk factors such as how fast we can run relative to the speed and distance of approaching vehicles. Also, we successfully ignore minor distractions when faced with the prospect of becoming a hood ornament on a late model Mercedes. We tend to concentrate our attention on this problem as we are acutely aware of the consequences of a mistaken calculation.

It is not a stretch to say that most humans are fortunate to muddle through modern life without the need to worry about physical threats to the same degree our ancestors did. That is partly because, for most of those reading this book, the conditions of modern living obviate the need to make the life-and-death assessments our ancestors confronted each day just to survive. Perhaps it is strange that we may not be as well equipped to assess risk associated with modern threats which seem inherently less serious.

Also, as a species we do not seem to be improving much in the risk management department, although evolutionary timescales are quite long so the jury may still be out. My guess is that the effect of evolutionary pressure on modern humans through natural selection has lessened based on our ability to overcome life-threatening hardships through advances in science, engineering, medicine, etc.

Notwithstanding their penchant for self-destruction and a seemingly endless capacity to inflict misery on each other, humans tend to be more rational than other species with which we share this planet. Before qualified academics attack my credibility for such a ridiculous statement, I do not mean that animals do not behave in effective ways that have evolved over time to increase their probability of survival. But through sheer brain power humans have created environments where their daily survival does not depend on a physical response to threats. We now confront risk problems of a different kind. These rely on analytical processes for which evolutionary pressures may not play as significant a role in the survival of the species.

Maybe we can turn to a slice of modern recreational life for insight into the more general problem of risk management. Examples of risk mitigation strategies can be found in baseball, America's national pastime. Endless repetitions of similar scenarios executed under highly controlled conditions provide mountains of historical data that have been used to develop effective methods of "survival" by winning teams.

For example, changing pitchers to counter a particular hitter or aligning fielders to adjust to a hitter's propensity to hit to one side of the field (recall the famous "Ted Williams shift") are examples of standard moves that might be witnessed in any game. Each move represents a calculated effort to strategically decrease the odds of your opponent scoring runs. Conversely, the team at bat has its own set of strategies intended to improve the odds of scoring runs. Both the offense and defense in baseball exploit a veritable avalanche of statistical data used to evaluate player performance and execute a team's respective strategies.

However, exclusively statistical analyses can lead to counterintuitive results. For example, an interesting analysis was conducted that utilized the full set of batting data from the inception of baseball to the present.[1] That study used a so-called Monte Carlo simulation to estimate the likelihood that a baseball hitting streak would exceed the current record of

56 games. The results were quite surprising. Fifty-six consecutive games by an individual player was not the longest expected streak, and Joe DiMaggio was not the likeliest of hitters to establish the record. The results of this simulation will be discussed in Chapter 4, since they are instructive on several levels.

Statistical representations are inherently generalizations, and probabilities do not imply certainty for a given situation. Each scenario is somewhat unique as players do not react the same way even under virtually identical conditions. Despite the volume of historical information, managers sometimes get burned when playing the odds and other times win big when relying on a hunch. For example, a manager might notice something tentative about a hitter that day or observe a flaw in a pitcher's mechanics and choose to ignore conventional wisdom. If things go well, the team's manager will be hailed as a genius in the next day's sports columns. If they do not go as hoped, the manager will be pilloried and too many gaffes of this kind will result in a precipitous loss of dugout privileges. Expectations and accountability run high in the transparent world of professional sports.

Statistical rigor alone is not always sufficient to minimize risk even in a highly structured world like baseball. So despite fantasies to the contrary, even the most intense math geek would not necessarily be a successful baseball manager. Consistently successful managers probably combine attention to statistics with leadership and intuition, the latter garnered through years of playing or managing. Paraphrasing a line from the film *From Russia with Love*: practice is nice but experience is everything. My contention is that analytical rigor and intuition based on experience are both relevant to evaluating risk. Furthermore, I suspect that individuals who are able to apply both in the correct proportion and at the right time are most likely to be successful in their respective endeavors.

I often successfully exploit intuition based on experience to make rough estimates of risk that affect my behavior. For example, when I go running in Central Park I am completely unfazed by the prospect of sharing the turf with the many squirrels that call this urban sanctuary home. Experience has taught me that the average squirrel would not attack humans and that it would be an extraordinarily hostile or demented little beast who would do so. Based on countless observations, I have informally concluded that squirrel behaviors are benign with respect to humans and a display of unprovoked aggression by a squirrel would represent an extremely rare event.

However, and this is an important point, even if I have misjudged the potential for an attack I am confident that I could fend off my assailant based on my modest size advantage. Therefore, in addition to an assessment of the potential for an attack as low, my vulnerability to injury is limited. I might add that I do not feel the same about large dogs or neurotic parents pushing strollers, so I assiduously avoid crossing their respective paths whenever possible.

It seems okay to grossly misjudge the likelihood component of risk if it can be compensated for by reducing the vulnerability to loss. But it is very important to correctly identify and understand the relative contributions of each component when assessing and mitigating risk, especially if your life depends on it. Surprisingly, even the smartest people get it wrong when leveraging personal experience to make judgments on risk; sometimes with significant consequences.

I observed a fantastic example of this while viewing a DVD on sharks. A prominent ichthyologist was demonstrating the indifference of bull sharks to humans as he and a journalist stood in waist-deep water with a school of these 500 lb. eating machines. As the sharks circled gracefully about their human hors d'oeuvres, the journalist nervously mentioned that he would only tempt fate in this way if accompanied by this world renowned scientist. This gives new meaning to the admonition "Don't try this at home" and is clearly nonsensical. When your plane is about to crash, it doesn't help that Chuck Yeager is sitting next to you . . . . unless he's flying the plane.

No sooner had the ichthyologist finished remarking how the sharks were oblivious to humans, presumably confirming his theory of shark behavior, when one of his heretofore indifferent swimming partners took a big bite out of his calf. In fact, the animal came close to dragging him off to deeper water to finish the job. It seems the ichthyologist may be a good biologist but is downright lousy at risk management.

The mistake here seems obvious to anyone who has ventured into water outside of a bathtub, but let's be precise. First, he misestimated the potential for bad things to happen even though he has probably observed sharks many times. My guess is that this gruesome event is not the statistical outlier in the same way as my fantasized encounter with a psychotic squirrel. What is absolutely certain is that the shark expert believed that a smallish likelihood component of risk would compensate for a rather significant vulnerability component; severe injury or death was always just one chomp away. The expert was supremely confident in his understanding of shark behavior and clearly assessed the potential for an attack as low.

Maybe the likelihood component of risk in this scenario is indeed low and the incident is a statistical outlier in the spectrum of possible shark behavioral outcomes (it would be difficult to do a simulation similar to the one analyzing hitting streaks in baseball). However, I would venture a guess that "average" bull shark behavior toward humans is considerably different than that of squirrels, and that even a small deviation from bull shark average behavior might be considered extremely aggressive. In statistical terms, the distribution of bull shark behaviors is quite narrow and peaks at about a relatively aggressive mean value. To put it more graphically, if our ichthyologist attempted this foolishness a million times there is a decent chance he would be missing a few body parts.

But the more salient problem from a security risk management perspective is to determine an acceptable estimate of the likelihood of an attack to off-set what amounts to nearly infinite vulnerability. For what it is worth, yours truly would want more than one expert's opinion before jumping into the water with a bull shark, especially since this species is historically the one that is most responsible for attacks on humans. The saying "If you swim with sharks you had better not bleed" does not guarantee safety under non-bleeding conditions.

Direct experience is indeed relevant to a security risk assessment process. Both experience and science are crucial to risk decisions in a world that seems to be neither completely deterministic nor entirely random. But rejecting science would be just as foolish as ignoring intuition developed through years of experience. Scientific reasoning and experience are not mutually exclusive in the world of security. Both should contribute to a rational assessment process that informs judgment in assessing the *totality* of risk.

This book focuses on the analytic processes necessary to assess risk and must assume the reader will develop the requisite intuition through his or her personal and professional experience. More specifically, it attempts to provide the security professional with the tools to analyze the individual components of risk and apply them to a structured assessment and mitigation processes. To do this, a firm understanding of threats and risk must be developed.

1.2 **THE FUNDAMENTAL EXPRESSION OF SECURITY RISK**

Threats are often narrowly considered to be forces or actions that cause harm. When pursuing a general analytic approach to risk, it is useful to think more broadly and consider a threat to be anything that has the potential to cause damage, loss, or worsening conditions. So, if state A describes conditions before the threat, and state B characterizes the state of play after the threat, state B is somehow "worse" than state A.

It is worth mentioning that although threats are not subject to moral interpretation, when analyzing risk polarity does make a difference. In other words, it matters if state B is better than state A or vice versa. However, this decision is subjective and may change depending on the view of the risk assessor. But the assessment and mitigation *processes* are frame of reference-independent. In other words, a person is free to determine what "worse" means and then create an appropriate risk model in accordance with inherently subjective ideas of good and bad.

People learn best by example, and it is important to supplement concepts with specifics anchored in everyday experience. Everyone would generally agree that events such as tornados, hurricanes, plane crashes, tsunamis, etc., are a threat in the classical sense since people, houses, and buildings are generally in worse physical shape after experiencing these events than before. Similarly, vehicle-borne explosives launched by terrorists are a no-brainer. But security professionals assess the risk associated with a variety of threats, which leads to a broad spectrum of potential scenarios.

In general it is not enough to merely identify threats and address them without regard to priority, especially when constrained by zero-sum budgets. Therefore, how do we determine which threats are worth fussing over and in what proportion? To answer this question and to develop an effective and proportionate mitigation strategy we must rigorously examine something called risk.

All security threats are characterized by a fundamental attribute called "risk," which consists of three components. In the absence of a threat there is no risk and there would be no need for security professionals. Although the notions of threat and risk are inexorably related, they are not equivalent. A threat cannot exist without the presence of individual risk components that in their totality characterize that threat. If any one of these components is missing, then by definition there is no threat.

Risk assessments must examine these constitutive elements, which are referred to as the individual components of risk throughout this book. These elements are (1) the potential or likelihood of threat occurrence, (2) the vulnerability to threat occurrence, and (3) the impact of threat occurrence. These components can be expressed in terms of what is referred to herein as the fundamental expression of risk. This is the underlying expression for all risk assessments:

$$\text{Security Risk (threat)} = \text{Likelihood} \times \text{Vulnerability} \times \text{Impact} \qquad (1.1)$$

This statement should be read as "The security risk associated with a threat equals the product of the likelihood of threat occurrence, the vulnerability to the threat occurring, and the impact should the threat occur." To accurately and completely assess risk it is necessary to evaluate each of these components. Mitigation strategies typically focus on the vulnerability component of risk simply because it is often the only component amenable to mitigation. Moreover, physical principles can sometimes be applied in assessing vulnerability and thereby facilitate a quantitative view on risk. This will be reviewed more in-depth in Part 2 of this book.

As noted previously, if any one of the components in the fundamental expression is zero then the risk is by definition equal to zero. In reality this never happens as there is always residual risk associated with any threat, although possibly not enough to warrant mitigation. This is a judgment call and represents the daily fare of security professionals.

This expression should not be interpreted as giving equal weight or importance to any individual component. For a given threat, the vulnerability to occurrence could be significantly greater than the likelihood of its occurrence or vice versa, even though the fundamental expression as previously stated appears to confer equal status to each component.

A good example of this might be the threat from terrorists in the form of a fission- or fusion-type nuclear device. The likelihood of occurrence is arguably quite low, but the vulnerability and impact (no pun intended) would be extremely high should such a device be detonated anywhere near your neighborhood. Realistically, there is no commercially viable mitigation against the effects of a fission- or fusion-based weapon detonated anywhere in proximity to a given facility.

Would it make a lot of sense for a security manager to attempt to protect a building from the effects of such a weapon? My guess is probably not since the expense associated with the mitigation strategy looms large in comparison to the potential for threat occurrence. Although the calculus of this trade-off is difficult to quantify and actually occurs quite frequently in security risk management, understanding the relative magnitude of each component of risk is ultimately at the heart of any rigorous assessment of risk and is what enables such decisions.

Much of this book is devoted to examining these components of risk in detail. However, one important point of clarification should be mentioned up front. Using the phrase "potential for occurrence" is sometimes more accurate than using "likelihood of occurrence" depending on the threat. The term "likelihood" is used rather liberally in the security vernacular, sometimes with unfortunate consequences. Security professionals are often asked about the likelihood of this or that incident occurring again. Consider terrorism, a threat that might be more correctly viewed as indiscriminate violence to further political or religious ends. What are the factors that contribute to the likelihood that this threat will occur against a particular target?

One can readily see that the factors affecting terrorism are not necessarily constant in time, especially if world politics is a significant inspiration for such incidents. Although coarse approximations can be made for specific locales, it is extremely difficult to specify the likelihood that a terrorist attack will occur with statistical confidence in a particular time interval or location. Historical data on previous occurrences may not be germane in estimating the likelihood component of risk, because the conditions driving such events sometimes change over even relatively short timescales. In such instances it is more precise to speak of the potential for a terrorist incident rather than likelihood and avoid the perception of a phony quantitative result.

The use of likelihood versus potential is more than a semantic distinction in this context. Suppose you are shown five playing cards face down. You have been told that one and only one of these is an ace. Then you are asked about the likelihood of selecting that ace from the group of five cards. You would have no trouble figuring the odds at 1 in 5 or 20%. Similarly, if it was known that a group of five identical buildings was targeted for a terrorist attack and the goal was to know the odds that a specific building would be chosen, the response would be the same as that given in the card game. If only it was this easy in view of the preponderance of rather speculative threat reporting.

In each of these examples, the spectrum of possible scenarios is well-prescribed, so calculating the likelihood of a specific scenario is straightforward. In a game of chance involving cards, all possible outcomes are known, although some fraction of the cards in a given hand is intentionally kept hidden to reduce an opponent's ability to perform a precise calculation of the odds. An assessment of this inherent uncertainty is the basis for each bet in a game of poker.

But recognize that this situation is entirely different from estimating where and when a security incident will occur. There may be an infinite number of possible venues and no time constraints imposed on an adversary. Even if it could be assumed that conditions will remain relatively stable, there is often a limited amount of historical data from which to estimate the likelihood of a future outcome. Statisticians refer to this condition as having a small sample space. Strictly speaking then, it is more appropriate to speak of a potential for occurrence rather than likelihood and to describe that potential in nonquantitative terms to avoid any hint of false precision.

This does not mean that a reasonable and sometimes even quantitative model of the risk associated security threats such as terrorism cannot be constructed depending on the approach. I will mention several commercially available models in Section 4.5 in Chapter 4, although the validity of these has never been rigorously tested (fortunately). In due course we will see that other security threats can be modeled in ways that are amenable to quantitative methods.

Physical principles often apply in assessing the vulnerability component of risk. One example of a physical quantity that can be used as a metric for the vulnerability to unauthorized signal detection leading to information loss is the signal-to-noise (S/N) ratio of an acoustic or electromagnetic signal containing confidential or sensitive information. A calculation of S/N as a function of distance from the radiating source yields the susceptibility to detection by competitors desirous of another company's information.

Another example is an estimation of potential damage incurred through detonation of vehicle-borne explosives as a function of payload and stand-off distance. Many other examples exist and span the space of modern security threats. Although vulnerability alone does not offer a complete picture of risk, approximations of relevant physical quantities using well-established physical models often yield insights into practical and effective mitigation strategies that can be used to estimate return on investment.

The impact component of risk is the most subjective and organization-specific of the three components. However, when conducting a risk assessment it is arguably the first component to consider. Big impact threats garner the most interest, and considering this component of risk up front will help prioritize efforts. Once it has been determined which threats would result in significant impact, additional analysis is required to determine the likelihood and vulnerability components of risk followed by a decision on whether or not to apply risk mitigation.

The flip side is that there is little need to worry about the likelihood or vulnerability for threats that have little or no impact on one's organization. It is for this reason that the impact component of risk is not addressed in detail in this book. It is presumed that those threats that have been determined to be of little or no impact to an organization are not worth analyzing further. Moreover, only someone familiar with an organization's inner workings could be expected to accurately assess the impact of a specific threat except in relatively obvious cases.

1.3 **INTRODUCTION TO SECURITY RISK MODELS AND SECURITY RISK MITIGATION**

Risk models represent an important tool in the quest to accurately assess the risk associated with threats and in applying effective risk mitigation. The word "model" can be confusing and possibly even intimidating, but it really represents a structured mechanism used to consistently delineate the risk associated with each unique threat. Well-constructed models actually facilitate strategic flexibility, a fact that belies its name. The creation of a risk model is a key step in the risk assessment process where the ultimate goal is to implement an effective and proportionate mitigation strategy.

The invocation of a model in any context does imply structure, which is definitely a characteristic we seek in analyzing security risk. Briefly stated, an immediate goal of the risk manager is to identify the set of unique threats and classify the risk factors associated with each of these in the most general terms. Categorizing the unique threats and their respective "food groups" of risk factors will facilitate an efficient application of mitigation methods without having to list every conceivable scenario. Once established, this methodology leads to consistency and coherence in the development of a risk mitigation strategy. These concepts including what is meant by the phrase "unique threats" will be explored in more detail in Chapter 3.

At a high level, model in security is no different in its form and function than the models used in other descriptive sciences. For example, biologists, botanists, pathologists, anthropologists, infectious disease specialists, psychologists, etc., all rely on classification schemes to characterize their respective objects of study. This enables them to identify patterns, groupings, and interesting deviations from the norm. Security specialists categorize threats and delineate the risk associated with those threats. They then develop mitigation strategies to counter those threats by addressing characteristics or features that enhance risk. I call these important features risk factors.

The medical profession offers insights into risk management. My father the pathologist would tell his medical students to merely describe what they see when examining a tissue specimen. Observing basic properties such as shape, texture, color, markings, etc., offers important clues to identifying disease. I recall him stressing that no medical expertise is required to perform this exercise. Because diseases cause repeated and classifiable changes to tissues, characterizing their physical properties is a crucial step in identifying the disease. The last thing you want to contract is a disease that doctors are incapable of classifying unless being the subject of a medical paper is more important than staying alive. Severe penalties for uniqueness are exacted in the world of medicine since identification and classification of a disease is usually a necessary antecedent to a cure.

When a patient complains of specific symptoms, a medical history is intended to correlate these symptoms with risk factors that enable the physician to generate a list of likely diseases. Family history, a recent travel itinerary, race, sex, religion, age, etc., are all features that contribute to an assessment of risk and the predisposition to specific diseases.

If given a choice, it is best not to wait until a disease is contracted before applying mitigation since the impact component of risk often worsens with time. To that end, prophylaxis in the form of inoculations is provided to travelers to reduce their vulnerability to diseases that are most likely to exist in specific destinations. Therefore, travel to such places is one risk factor associated with specific infectious diseases. The use of prescription medicines in general represents one form of risk mitigation to counter the threat of disease.

In the world of security, threats are analogous to disease in medicine. Patients are the target of the disease and a regimen of drugs, surgery, lifestyle changes, and/or physical therapy in some combination as dictated by experience, science, and/or judgment comprises the risk mitigation strategy. Tragically, there was a rationale, however distorted, as to why the World Trade Center was targeted twice by extremists. The iconic nature of that particular building made it an attractive target to anti-Western groups enhancing the likelihood component of risk. Specific features of that facility also affected the vulnerability to attack and so influenced that component of risk. These features are the risk factors. Understanding and classifying the risk factors associated with each component of risk, which is the crux of a risk model, are essential to developing an effective mitigation strategy for any risk scenario.

To illustrate the concept of a risk model more completely, let's examine the threat associated with visitors to a facility. The risk factors associated with this threat are (1) an increased vulnerability to incidents of theft, sabotage, and/or information loss due to the physical proximity to sensitive internal areas and (2) an enhanced potential for such incidents based on the inherent uncertainty associated with individuals with whom a company is unfamiliar.

These risk factors apply to both the pizza delivery guy and to the visiting CEO. Here is where judgment is required to develop a mitigation strategy that is both sensible and proportionate to the risk associated. And this is precisely what security directors are paid to do. For example, one might assume that greater levels of scrutiny are required for the pizza delivery guy than the visiting CEO. But this might not be correct, especially if the visiting CEO is a fierce competitor. The challenge is to develop a broadly applicable model that accurately characterizes the risk and facilitates effective mitigation without compromising the company's business model or violating its culture. In the case of the threat from visitors, the risk model should lead to rigorous but flexible authorization and authentication procedures.

Stated once more for emphasis, the process of developing a risk mitigation strategy begins with identifying the set of unique threats and associated risk factors for each of those threats. Although this sounds easy, it can often be a nontrivial exercise because it requires a very thorough understanding of the set of unique and relevant threats. For example, there is sometimes a temptation to develop a list of specific scenarios and address each individually rather than think in terms of those broad "food groups" of threat types.

This is both inefficient and potentially less informative since invariably certain threat scenarios can be shown to be roughly equivalent and therefore grouped together. Each group is then effectively addressed by similar mitigation methods. Listing all scenarios invites complexity and can lead to confusing, duplicative, and/or wasteful mitigation strategies. My contention is that such an approach is more likely to result in ignoring important threat types.

Such an approach would be equivalent to studying insects by attempting to describe each bug individually rather than identifying general categories of bugs based on common features. In addition to wasting a lot of time and effort, such a methodology would make it difficult to see patterns and recognize an evolutionary trend or identify various subspecies. The search for common themes and patterns applies equally to the world of insects and security risk.

Paraphrasing 4th century BC philosopher Kuan Tsi as recounted in the book *Axis and Circumference*,[2] a beautiful treatment of the role of classification and scale in the plant world:

> *Reality is the embodiment of structure;*
>
> *Structures are the embodiment of properties;*
>
> *Properties are the embodiment of harmony;*
>
> *Harmony is the embodiment of congruity.*

Continuing with a quote from the author, Steven Wainwright:

The way to breathe life into the description of any object is to apply adjectives to it. A piece of cloth has little interest for us until we know whether it is starched, hand woven, salmon pink, translucent, knotted, torn, bespangled, or sodden. Properties are adjectival and adverbial descriptors that link structures and materials to specific functions.

Ultimately, the quest of the security professional is to understand the spectrum of unique threats and to accurately classify them pursuant to assessing their individual components of risk and then applying mitigation. The remainder of this book is focused on providing the tools to enable the security professional to do just that.

1.4 **SUMMARY**

The concepts of threat and risk are different albeit profoundly interrelated. The risk associated with a threat represents its defining attributes. A complete characterization of the individual components of risk is required to establish an effective mitigation strategy. The fundamental expression of risk can be expressed as follows:

$$\text{Security Risk (threat)} = \text{Likelihood} \times \text{Vulnerability} \times \text{Impact} \qquad (1.2)$$

It pays to retain the image of the bull shark and the tasty ichthyologist to help keep the individual components of risk in focus and be alert to unhealthy trade-offs when developing a risk mitigation strategy.

The important notion of a risk model was introduced in this chapter. This is a delineation of the specific factors that influence one or more components of risk for each unique threat. Such models can be developed for threats affecting facilities or security-related processes such as mail screening, visitor authorization and authentication, etc.

Although understanding the limiting cases of risk is important, extreme scenarios are not always particularly useful in terms of a mitigation strategy. For example, a vehicle-borne explosive detonated a few feet from an office building will most likely cause a bad day for everyone in the vicinity. But developing a totally effective mitigation strategy for this threat in the face of the realities of urban life may be unrealistic. Assessing security risk and applying mitigation must be based on rigor but balanced by judgment, reason, available resources, and other eminently practical considerations.

REFERENCES

1. Arbesman S, Strogatz SH. A Journey to Baseball's Alternate Universe. *New York Times Op Ed*. March 30, 2008.
2. Wainwright SA. *Axis and Circumference: The Cylindrical Shape of Plants and Animals*. iUniverse; 1999.

The fundamentals of security risk measurements

The more complex the mind, the greater the need for simplicity of play. — Captain Kirk

"Shore Leave," *Star Trek,* **Stardate 3025.8**

2.1 **INTRODUCTION**

Rigorously assessing security risk often involves understanding how certain physical quantities are affected by changes in parameters like time or distance. Physical quantities such as intensity and concentration are important concepts in security risk, and these underscore the inherently technical nature of security risk. Moreover, certain mathematical functions and representations are useful in characterizing these quantities and these are introduced in this chapter. The good news is that the breadth of terms, concepts, and elementary physical models needed to develop ballpark estimates of risk is relatively limited. These appear repeatedly in Part 2 of this book, which examines the likelihood and vulnerability components of risk in greater detail.

2.2 **LINEARITY AND NONLINEARITY**

Risk can sometimes be evaluated more precisely because of its connection to relevant physical quantities that obey natural laws. Such laws describe how these quantities change as a function of scenario-dependent parameters like distance and time. The effect of the behavior of these quantities on risk using a range of parameter values dictated by scenario-specific conditions can then be examined.

For example, if the hope is to evaluate the vulnerability of a facility to vehicle-borne explosives, then it must be done for various payloads and standoff distances to develop an effective mitigation strategy. It is unlikely an adversary will advertise the plan of attack; therefore a distribution of attack scenarios is required. Because an estimate of the vulnerability to damage is related to physical quantities such as explosive impulse, over-pressure, etc., it is important to accurately determine their dependence on parameters such as the distance from the source to the target and the explosive payload. This process of calculating the behavior of physical quantities as a function of scenario-dependent parameters can be a main-stay of assessing the vulnerability component of risk.

Moreover, understanding the resultant *change* in risk as these scenario-dependent parameters vary is essential in developing a risk mitigation strategy. Perhaps it is surprising that knowing the shape of a curve that graphically expresses this change is this important to understanding risk.

Several key concepts in thinking about risk, and specifically about the vul-nerability component of risk, are the notions of linearity and nonlinearity. The term linearity is rooted in the word "line." If two variables are linearly related it means a change in one variable causes a proportionate change in the other. Equation 2.1 is a made-up expression that depicts a hypothe-sized relationship between chronological age and body weight:

$$\text{Body Weight} = (4 \times \text{Age}) + 25 \tag{2.1}$$

Based on this equation, we can calculate body weight by simply substituting representative values for age:

Age (years)	Body Weight (pounds)
10	65
20	105
30	145
40	185
50	225

In this case age and body weight are linearly related because an increase in age causes a *proportionate* increase in body weight for all values of age. Specifically, an increase in age from 10 to 20 years yields a weight differ-ence of $105 - 65 = 40$ lb. If one doubles the age difference from 20 to 40 years, this yields a weight difference of $185 - 105 = 40 \times 2 = 80$ lb. In other words, doubling the difference in age results in a doubling of the difference in weight.

This fixed relationship between differences in related variables is the essence of linearity. The fact that changes in one variable always produce

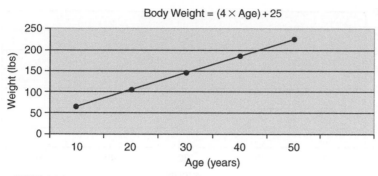

FIGURE 2.1 Linearity.

proportionate changes in the related variable throughout the range of the function is the key element here. A graph of the expression weight = (4 × age) + 25 is depicted in Figure 2.1, and it is worth noting the very straight and hence *linear* appearance. Linearity is an important characteristic of physical quantities and associated parameters that affect risk, since variables that are linearly related change in relatively manageable increments.

However, the physical world is replete with examples of processes or quantities that display *nonlinear* behavior. A nonlinear relationship between variables or parameters is one where successive changes in one variable cause *disproportionate* changes in a related variable. An example of a nonlinear relationship is the velocity of an object and its kinetic energy. The kinetic energy of an object is proportional to the square of its velocity. The exact expression is $KE = \frac{1}{2} mv^2$ where m is mass. Let's plot some representative values of kinetic energy in terms of velocity without specifying any particular units (I assume the mass is one or unity):

Velocity	Energy
10	50
20	200
30	450
40	800
50	1250

If we double the velocity from 10 to 20 this results in a difference of 200 − 50 = 150 units in energy. However, doubling the velocity once more from 20 to 40 units yields a change in value of 800 − 200 = 600 energy units. We see that a doubling of the velocity results in a quadrupling of the kinetic energy. This nonproportionate relationship between the related

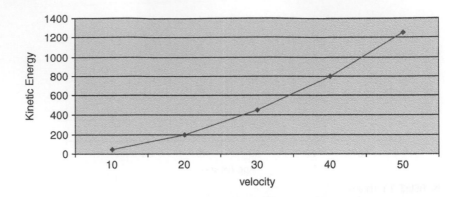

■ **FIGURE 2.2** Nonlinear function: $KE = \frac{1}{2}mv^2$.

quantities, in this case energy and velocity, is the defining characteristic of nonlinear functions. As the difference in the values of velocity increases, the difference in the kinetic energy becomes disproportionately larger. A graph of the expression for kinetic energy is $KE = \frac{1}{2}mv^2$ using the values shown in Figure 2.2. Notice its non-straight-looking appearance:

This is a good time to point out an important feature of the expression for kinetic energy. The exponent of the velocity variable indicates the number of times that variable is multiplied by itself. In other words, for any variable x, $x^2 =$ x times x, $x^3 =$ x times x times x, etc. In particular, an exponent of two is referred to as the square of a variable, so $x^2 =$ x squared. In the same vein, a variable with an exponent of three as in x^3 is referred to as x "cubed." x^4 is referred to as "x to the fourth power" and x^5 is "x to the fifth," and it continues ad infinitum.

A discussion of the mechanics of exponents is an excellent segue into how these mathematical expressions are relevant to characterizing security risk. In Section 2.5, we will see how the concept of scale, which directly relates to exponents, plays an important role in this regard. Linear and nonlinear functions tell an important story about how physical quantities and scenario-dependent parameters change with respect to one another. In understanding risk it is sometimes more relevant to know how a relevant physical quantity such as intensity, signal/noise (S/N), etc., changes with angle, distance, or time, than it is to know the magnitude of that quantity at a specific point in space or time. The range of possible threat scenarios varies considerably, so it is important to use all appropriate information to estimate the limits of vulnerability associated with relevant scenarios and plan accordingly.

Security strategies often call for an explicit relationship between a physical quantity affecting risk and a dependent parameter. Examples might include the intensity of a signal as a function of distance from the signal source or the volumetric density of a chemical agent as a function of time.

Once this relationship is understood, a graphical representation of the dependence can be useful in developing a physically realistic mitigation strategy.

For example, it might be important to know how the sound intensity changes from one side of a wall to the other as a function of the thickness of the wall material when designing a conference room. Or maybe it is required to determine the intensity of radioactivity from a radioisotope source as a function of distance from that source to determine the characteristics of a proposed shield. There are many other examples, and some of these are discussed in more detail in Part 2 of this book.

When armed with such physical insights, effective mitigation strategies are more likely to emerge. The shape of the curve depicting the intensity versus distance from a source emitting some form of nasty energy yields important information regarding security risk, which in turn facilitates effective and proportionate mitigation. It is difficult to imagine a complete picture of risk for many physical security threats without such a model.

This book discusses a number of security-related processes that are characterized by nonlinear dependencies on physical quantities. One example is the intensity of sound with respect to distance from a point source of acoustic energy (e.g., a person from the source of energy speaking). Here the intensity falls off as the inverse square of the distance r or $1/r^2 = r^{-2}$. The exponent of 2 indicates a nonlinear relationship between intensity and distance. The minus sign in the exponent implies that the intensity *decreases* as the distance from the audio source increases. The implication is that disproportionate effects will accrue from changing the distance to sources of radiating energy, and this impacts the vulnerability component of risk and quite possibly the resulting mitigation strategy.

But a function of the form $y = x^2$ such as in the equation characterizing kinetic energy is not the only nonlinear expression. It is important because so-called quadratic equations occur frequently in the physical world. An exponent can assume any value. A physical quantity that contributes to the effect of explosive blasts on structures is the so-called impulse. This characterizes how long the pressure wave interacts with a structure and is measured in units of pressure multiplied by time (i.e., pounds per square inch-millisecond or psi-ms). For those schooled in physics, the impulse represents the time integral of the momentum imparted to the structure by the explosive force. I will revisit this concept later in Chapters 5 and 6.

According to one reference cited in Section 5.2.1, the impulse varies inversely with distance from the source (i.e., as $1/r = r^{-1}$). The same reference specifies that the overpressure scales as one over the distance cubed with distance from the source.

Therefore, the effect of increases in the distance on explosive impulse is considerably less than its effect on overexposure. Understanding the dependence of a physical quantity on a parameter such as distance is often key to understanding the vulnerability component of risk.

Representations of impulse and overpressure as a function of distance from an explosive source are examples of decreasing nonlinear functions. Several "generic" decreasing nonlinear functions are shown in Figure 2.3, where the function f(r) is plotted relative to the variable r. We see that f(r) approaches zero as r becomes larger in value but never actually equals zero. Mathematicians refer to the behavior of functions like f(r) as *asymptotically* approaching a limiting value.

We could just as easily encounter nonlinear functions that *increase* with distance, time, etc. Two examples, $y = x^2$ and $y = x^3$, are plotted in Figure 2.4:

■ **FIGURE 2.3** Decreasing nonlinear functions.

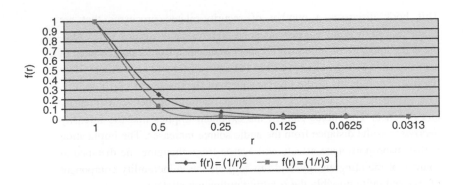

■ **FIGURE 2.4** Increasing nonlinear functions.

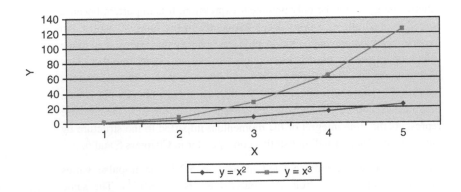

2.3 **EXPONENTS, LOGARITHMS, AND SENSITIVITY TO CHANGE**

Scenario-dependent parameters are sensitive to change; linear and nonlinear expressions that describe the behavior of physical quantities affecting security risk can provide insight into this behavior. Here are a few examples of such expressions using the exponential shorthand which demonstrates this sensitivity firsthand.

We now know that $r^3 = r \times r \times r$, $r^6 = r \times r \times r \times r \times r \times r$, and $r^9 = r \times r \times r \times r \times r \times r \times r \times r \times r$, and so on. According to the definition of exponents, the number of "r"s on the right-hand side of the equation corresponds exactly to the superscript above the r on the left side of the equation. If we substitute the number 10 for r, the value of these functions is a thousand, a million, and a billion, respectively (i.e., $10 \times 10 \times 10 = 10^3$, etc.). Clearly small differences in the exponent can make a big difference in the magnitude of the function.

A personal story may serve to drive this point home. I once deposited a check for $10,000 in my bank account. When I checked my balance, the bank had credited my account for only $1000. The $9000 discrepancy was reflected in the omission of a single zero. This was equivalent to reducing the exponent of 10^4 (i.e., 10,000) by one to 10^3 (i.e., 1000). I was ultimately reimbursed in full after producing my deposit slip. So the morals of this story are to keep all deposit slips and to pay close attention to the exponents in all transactions.

Speaking of money, exponents play an important role in the financial world. Each of us with an interest-bearing bank account is affected by the time period over which interest on the principal is calculated. The general formula is Future Value = (Present Value) $\times (1 + I/n)^n$, where I is interest rate, and n is the number of times interest is applied to the principal. Consider the difference in the value of your bank account after one year if the interest rate was applied once per day versus once per year using the same interest rate. That translates to the exponent n equaling 365 in the former and 1 in the latter case. As I write this I am lamenting the pathetic rate of return I am experiencing in my conventional savings account.

Exponents have a powerful effect on the range of values of a function and tell an important story about the processes they characterize. It is the exponent that determines the degree of nonlinearity and hence how rapidly a quantity changes as a function of a specific parameter. I pointed out how the sound intensity, I, changes as a function of distance, r, from a point source of acoustic energy. Such a calculation could yield valuable information on the vulnerability to conversation-level overhears at every point evaluated. The relationship is compactly written in the form $I = r^n$ where the exponent, $n = -2$ determines the sensitivity of the intensity to changes in the distance from the source, r.

The so-called logarithm is the inverse of exponentiation. It therefore "undoes" the effect of the exponent. For example, the logarithm of 1000 in base 10 is the value of n in the expression $10^n = 1000$. In this case n = 3. Similarly, the logarithm (often abbreviated as "log") of 100 equals 2, and the log of 1,000,000 equals 6, etc. A list of exponents in base 10 and their integer equivalent expressions is

$$10^0 = 1$$
$$10^1 = 10$$
$$10^2 = 100$$
$$10^3 = 1000$$
$$10^4 = 10,000$$
$$10^5 = 100,000$$
$$10^6 = 1,000,000$$
$$10^7 = 10,000,000$$
$$10^8 = 100,000,000$$
$$10^9 = 1,000,000,000$$
$$10^{10} = 10,000,000,000$$

By the way, the previously esoteric mathematical term "google" that is now familiar to nearly everyone in the world is defined as 10^{100}.

The logarithm is important in presenting information about risk for several reasons. First, it represents a compact way to display the full range of values when plotting a broad range of numbers. Consider a graph of $y = x^3$, for x = 1 through x = 100. The corresponding values of y range from 1 through 1,000,000. If we plot the logarithm of y instead of y, the y-axis extends from 0 to 6. Figure 2.5 shows a plot of $y = x^3$ for x = 1 to 100 using a semi-logarithmic (i.e., only one axis is logarithmic) scale. Imagine the size of the graph that would be needed if we plotted all values of y from one to one million.

Furthermore, sometimes functions describing physical processes vary as the logarithm of a parameter rather than the linear value of the parameter. Examples of this abound in the physical world. Processes or quantities that vary as log(x) have a more gradual dependence on x than those that scale linearly with x.

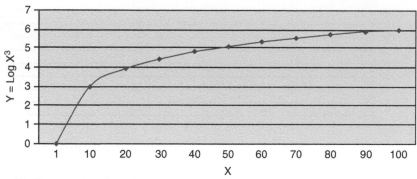

■ **FIGURE 2.5** Semi-logarithmic plot.

2.4 **THE EXPONENTIAL FUNCTION** e^x

For some reason certain numbers and expressions occur more frequently than others in nature. One of those is the number "e" and is defined by the following expression (written to only four decimal places, e is a so-called irrational number and does not have a finite decimal representation), $e = (1 + 1/n)^n = 2.7183 \ldots$ as n approaches infinity.

The exponential function is defined as $f(x) = e^x$ where x is some arbitrary exponent. This function appears in many contexts in the natural world and is the solution to a number of equations related to physical processes that characterize risk. The logarithm in base e has a special designation and is referred to as the "natural logarithm." This does not imply that the regular logarithm is in some way unnatural. The natural logarithm is abbreviated as "ln." We now know from the previous discussion on logarithms that $\ln (e^x) = x$. This is because the natural logarithm undoes exponentiation for functions where e is the base in exactly the same way that regular logarithms do for expressions in base 10 or any other base.

Figure 2.6 shows a plot of $y = e^x$ where the exponent varies from $x = 1$ to $x = 5$. It is immediately obvious from the graph that this is a nonlinear function. The exponential function is a solution to a common and relatively simple differential equation that governs a number of processes relevant to physical security. This statement applies equally to the *decreasing* exponential function $y = e^{-x}$ depicted in Figure 2.7. Because of its general applicability to physical processes related to security, this function will appear throughout the book. Understanding the general behavior of the

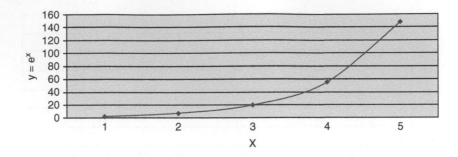

■ **FIGURE** 2.6 Increasing exponential function.

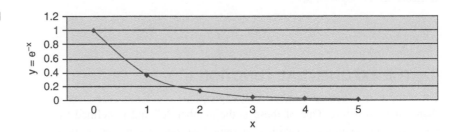

■ **FIGURE** 2.7 Decreasing exponential function.

exponential function can lead to important insights into the vulnerability component of risk.

2.5 **THE DECIBEL**

The decibel (dB) is very useful in depicting relative changes in quantities related to security. Decibels can be intimidating, especially if logarithms are scary since the logarithm is an inherent part of the definition of the decibel. Simply put, a decibel is merely one means of expressing a ratio between two quantities. The decibel indicates the magnitude of one quantity relative to another expressed in logarithmic units.

It makes no sense for individuals to say a signal is 30, 40, and 100 dB, although such statements are made all the time. A response to this statement must be "30 dB compared to what?" You may have seen sound pressure levels seemingly expressed without such a comparative reference. For example, conversation-level speech is sometimes expressed as 60 dB. In this case decibels are actually referenced to the minimum threshold of human hearing (10^{-12} W/m^2), and this figure forms the basis for the Sound Pressure Level (SPL) scale. The full expression is 60 dB SPL, which is consistent with the definition of the decibel as a unit expressing *relative* magnitudes.

The strict definition of a decibel when used to compare quantities like power or intensity is 10 log(x/y). We could just as easily be comparing the value of Joe versus Tom's bank account or any comparison for that matter. For example, if I happen to mention that the magnitude of my affinity for vanilla ice cream is twice that of chocolate, and we let x = the magnitude of my taste for vanilla ice cream and y = the magnitude of my taste for chocolate ice cream, then the ratio of magnitudes = (magnitude of x)/(magnitude of y) = 2.

Plugging that into the definition of a decibel we get 10 log(2) = 3, since the logarithm of 2 in base 10 is 0.3. Therefore, an equivalent statement for my liking vanilla ice cream twice as much as chocolate is that my fondness for vanilla is 3 dB more than chocolate. If my taste for vanilla ice cream actually exceeded that of chocolate ice cream by a factor of 1000 then 10 log(1000) = 30. In this case my lust for vanilla exceeds that of chocolate by 30 dB. It is really that simple and once you become accustomed to thinking in terms of decibels it becomes second nature.

If we want to convert 30 dB back to its equivalent linear expression we must first divide 30 by 10. This "undoes" the multiplicative factor of 10. This act of undoing yields the number 3. Now we must deal with the logarithm portion of the expression. Exponentiation undoes the logarithm function, so if we operate in reverse and raise 10 to the third power (i.e., 10^3) we restore the original linear value of 1000.

Decibels are sometimes expressed in terms of a specific physical quantity for reference. Examples include decibels relative to a milliwatt (dBm) and decibels relative to a volt (dBV). A 0 dBm signal is 1 mW(i.e., one thousandth of a watt); a 30 dBm signal is therefore equal to 30 dB (i.e., a factor of 1000) above 1 mW or 1 W.

Also, I noted above that a decibel often compares quantities like power and intensity and is given by 10 log(x/y). When comparing amplitudes it is convenient to compare the square of the respective quantities. So dB = $10 \log(A^2_1/A^2_2) = 20 \log(A_1/A_2)$ based on the properties of logarithms.

You now know most of what you will ever need to know about decibels. Let's work a real-world security example that will be explored in more detail in Chapter 5 where the vulnerability to detection of radio frequency signals is discussed.

Suppose you were interested in understanding the vulnerability to an adversary detecting a radio frequency signal broadcast at a frequency of 10 MHz by a wireless headset. This entails calculating how high the received signal intensity is above the minimum threshold, dictated by the thermal noise power or kT_0B (i.e., the so-called Johnson noise) for the electronic components in an adversary's antenna and receiver. Here k is Boltzmann's constant, T_0 at room temperature is the absolute temperature $= 293$ K, and B is the signal bandwidth.

It turns out that kT_0B for a 1 MHz bandwidth signal, B, is equal to 0.04×10^{-13} W. We need to know what the noise threshold is at the specific broadcast frequency to figure out the minimum detectable signal by an adversary. Let's assume your facility is located in a rural area and that man-made noise, the principal noise source at 10 MHz, is known to be 40 dB *relative to* the minimum noise signal, kT_0B.

This means the dominant source of external noise is a factor of 10,000 above the thermal noise limit, previously calculated to be 0.04×10^{-13} W. So, the resulting noise floor is actually $10^4 \times 0.04 \times 10^{-13} = 0.04 \times 10^{-9}$ W. And, if we say a signal must be a factor of 10 above the actual noise floor to be detectable, then that signal must be $10 \times 0.04 \times 10^{-9} = 0.4 \times 10^{-9}$ W to be vulnerable to unauthorized detection. Figure 5.5, which specifies ambient noise power relative to kT_0B (in decibels) in various frequency regimens, relies on this exact calculation to determine the ambient noise as a function of frequency.

If all this seems somewhat confusing, then it may be helpful to merely memorize some common decibel values. You have already seen that 3 dB is a factor of 2. Continuing, 10 dB is roughly a factor of 10, 20 dB is a factor of 100, 30 dB a thousand, and 60 dB represents a factor of one million. Do you detect a pattern here? Each increase of 10 dB represents a factor of 10 increase in the equivalent linear expression. One of the really nice things about decibels is that they can be added together obviating the need to multiply large numbers. This convenient feature is derived from defining the decibel in terms of a logarithm.

For example, if we know that each acoustic barrier offers 30 dB $= 10 \log$ 1000 or a factor of 1000 of sound attenuation from one side of the barrier to the other, and there are three identical barriers between the source and listener, it is easy to determine the total attenuation. This is accomplished by simply adding 30 dB $+$ 30 db $+$ 30 dB $= 90$ dB $= 10 \log(10^9)$ or a total attenuation factor of one billion. The alternative method would require multiplying each factor, i.e., $1000 \times 1000 \times 1000 = 1,000,000,000 = 10^9$, which can be more cumbersome. Hopefully it is now clear why expressing numeric comparisons as decibels is so convenient and so often used.

2.6 **SECURITY RISK AND THE CONCEPT OF SCALE**

Identifying the physical quantities that affect security risk is of great importance in understanding and mitigating threats. It is also a central focus of this book. Once these quantities have been identified, a key step is to examine their sensitivity to change *or how they scale* relative to the change in the value of scenario-dependent parameters.

Specifically, the magnitude of physical quantities that affects security risk can change as a function of familiar parameters like distance or time. For example, one may want to know the intensity of electromagnetic or acoustic energy as a function of distance from the source since this could be important in understanding the vulnerability to information loss. To a first approximation, understanding whether this relationship is linear or nonlinear will be very helpful in managing the risk associated with the specific threat of concern.

However, usually a more detailed specification is required to fully appreciate the vulnerability component of risk. In other words, a more precise accounting of how intensity scales with distance is needed. Let's examine this point in more detail. I note in passing that the American Museum of Natural History in New York has a permanent exhibit in the planetarium that beautifully illustrates the concept of scale in the physical universe.

The expression that characterizes the intensity of isotropically radiating energy sources as a function of distance from the source is $I = I_0 \, (r_0/r)^2$. I is the intensity at any point in space, I_0 is the initial intensity at the distance r_0, and r represents an arbitrary distance from the source. Don't worry about the physical details right now as I will discuss intensity more precisely in Chapter 5. The exponent of r in this case is -2 (i.e., r is in the denominator), signifying an inverse square relationship between I and r. In other words, the intensity *decreases* with *increasing* distance from the source.

Recall that the exponent is the number that characterizes the sensitivity of a quantity, in this case intensity, to changes in a parameter such as distance. The exponent can therefore be a key feature in characterizing the vulnerability component of risk. This is because the intensity of the sound that impacts someone overhearing a conversation can now be evaluated at *any* distance from the source.

The intensity of a source of energy as a function of distance is not the only relevant physical quantity to security risk. But it is very important and occurs quite frequently in scenarios of interest since detectors are sensitive to energy intensity. The vulnerability component of security risk is sometimes related to physical quantities whose magnitude scales with time and/or distance. Indeed, one might also want to know how quantities like intensity and concentration are affected by parameters such as frequency, energy, angle, etc., depending on the scenario.

Generic inverse square and inverse cube functions were shown in Figure 2.3. One readily observes that $y = r^{-3}$ decreases more rapidly with changes in r than $y = r^{-2}$. To put things in security-related terms, if r corresponds to distance and y represents the intensity of a signal carrying sensitive information, there would be *less* vulnerability to compromise for those signals that decrease *more* rapidly with distance. Such knowledge would be important in developing a mitigation strategy to ensure sufficient standoff exists between the signal source and someone intent on unauthorized detection.

Understanding how quantities scale with scenario-dependent parameters can offer useful insight into physical processes related to security risk and even sometimes help explain counterintuitive phenomena. In an example that has nothing to do with security, smaller and presumably "weaker" people can typically do more pull-ups than their bigger and "stronger" counterparts. Why is this? The idea of "smaller is stronger" has particular appeal to yours truly who is probably several standard deviations smaller than the average American male.

Consider the effort involved in a pull-up. A person hangs vertically from a bar and pulls his entire body upward until the chin is over the bar. Upper body muscles work against the force of gravity that is exerted by the earth on the entire body. Muscular strength is related to the cross section of the muscles involved so that the greater the cross section of the muscle, the more force can be generated.

It is true that more muscular individuals can apply more force in a one-time demonstration of work, where work is equal to the force times the distance over which the force is applied. Large-muscled people are generally bigger and weigh more than those of the smaller variety. The problem for big people is that muscle cross section only scales with muscle area or length squared. However, body mass and the force required to lift that mass (i.e., a person's weight) is proportional to volume or length *cubed*. Therefore, bigger individuals "lose" according to the ratio of length squared to length cubed with respect to the work required to do multiple pull-ups.

Pull-ups are a true test of muscular endurance and this scaling argument shows why less hefty individuals actually have an advantage in repetitive movements of body mass. Apart from deriving perverse pleasure in flaunting a modest competitive advantage over my more macho colleagues, this lesson illustrates how scaling can help explain physical phenomena. In exact analogy, the magnitude of physical processes related to security sometimes changes at a rate governed by physical parameters that can be described by a single exponent. In these instances assessing the vulnerability component of risk can depend on discovering this exponent.

In summary, three questions should be asked up front when performing a risk analysis for physical threats: (1) what are the relevant physical processes or quantities affecting the individual components of risk, (2) what parameters influence the magnitude and/or rate of change of these quantities, and (3) how do these quantities scale with the aforementioned parameters. Ideally, one develops an intuition about such things over time. Much more will be said about this approach to risk analysis in future chapters as this represents a fundamental theme in the risk assessment process.

2.7 SOME COMMON PHYSICAL MODELS IN SECURITY RISK

Security scenarios and estimates of risk often involve analyses of physical quantities that are functions of distance and time. Such quantities characterize the magnitude of a threat in time and/or space. Mitigation is usually accomplished by getting as far from a source as possible, waiting a sufficient amount of time until the threat abates, or installing physical barriers of adequate thickness to protect against detection or other unhealthy consequences. Some representative scenarios include the effect of a bomb exploding, a radiological source emitting gamma radiation, contaminated air leaking into a building, etc.

It would be hopelessly naive to suggest that all physical processes that impact security could be summarized in one section of one chapter of a single book. In-depth analyses of physical phenomena that relate to air flow, explosive reactions, etc., are complicated and precise answers require detailed analyses and/or computer modeling. However, there are a few simple physical models that are encountered frequently when evaluating security scenarios. These can yield a rough, but nonetheless insightful, appreciation of risk for a broad range of security risk scenarios.

Let's examine several physical quantities that arise frequently in physical security, although much more will be said about these in Chapters 5 and 6. The notion of the concentration of a substance is ubiquitous in biological and chemical scenarios. Deleterious effects often depend on how concentrated the substance is in air, since we spend most of our lives immersed in air and need it to breathe. In other words, we often want to figure out the ratio of the quantity of some nasty substance to a specified volume of air. Concentration is often expressed in grams of a substance per liter or milligrams per cubic centimeter, etc. Irrespective of the substance, concentration in this case refers to a ratio of mass to volume.

The concept of density is also important in understanding phenomena affecting physical security scenarios. We use the term density frequently in our daily discourse; all the more reason to ensure the meaning is precise. Densities indicate how much of a substance or any "thing" exists within a specified portion of space. For example, the concentration of a substance in a liquid expresses density in three dimensions: linear, areal, or volumetric.

A number of security scenarios exist where the density of some nasty stuff is spread out over a given area and the effect on risk is assessed. One example that will be discussed in Chapter 5 relates to radioactive dispersion devices where areal density of radioactive material is important in assessing the vulnerability component of risk to health.

I have already commented several times on the importance of intensity in assessing security risk. Intensity is an inherently dynamic physical quantity since it is parameterized in terms of time. In fact, intensity combines the parameter of time with density since it specifies the rate of energy *flux* or equivalently as power density. In the case of sound, it is the total amount of energy associated with oscillations of a medium (often air) transferred during each second through a surface of unit area and perpendicular to the direction of propagation. Intensity is often measured in terms of W/m^2 or W/cm^2, etc., and appears in optics, electromagnetism, acoustics, or any phenomenon where there is energy propagation.

There are many more physical quantities that are relevant to security such as pressure (force/area), velocity (distance/time), acceleration, (velocity/time), etc. These quantities will be encountered in the course of analyzing a broad range of security problems. However, there are also a surprising number of security risk scenarios that can be described on a high level by a relatively small set of physical models. I believe these are worthy of further discussion and in-depth understanding because of their general applicability to assessing risk.

A common security scenario is that of a localized or point source radiating some kind of energy. Familiar examples of point sources include a linear antenna radiating an electromagnetic signal, a small chunk of radioactive material giving off gamma radiation, or a human mouth generating acoustic energy otherwise known as speech. In general, a point energy source is one where the physical size of the source is smaller than the wavelength of the radiated energy or small compared to the distance from the source.

You may recall from your days in geometry class that the surface area of a sphere is $4\pi r^2$ where r is the radius of the sphere. If the intensity is I at a distance r from the source, then at a distance 2r the intensity is reduced by $1/(2)^2$. I noted previously that in this case there is said to be an inverse square relationship of intensity as a function of distance from the source. For a signal radiating with power P, its intensity at any distance r from the source of radiation can be written as

$$I = P/4\pi r^2 \tag{2.2}$$

It is the intensity of a source of energy that is often the most useful physical parameter in assessing vulnerability, since it is the time rate of energy deposited over a given area that is detected by a sensor and is therefore most impactful. Figure 2.8 illustrates the situation for a point source of radiating energy.

We see from this figure that a point source radiates an ever-expanding "balloon" of power, so that the intensity or power per area of surface becomes more dilute at greater radii from the source. This is a common theme in physical security scenarios. A plot of the intensity versus distance would look like the $1/r^2$ function depicted in Figure 2.3.

If the source of energy is a chunk of radioactive material, the intensity of the radiated energy is a key factor in assessing the vulnerability to radiation exposure. If the source of energy is a linear antenna or someone's mouth, the intensity is also important in determining whether sensors deployed by an adversary can detect the signal hence providing insight into the risk of unauthorized signal detection.

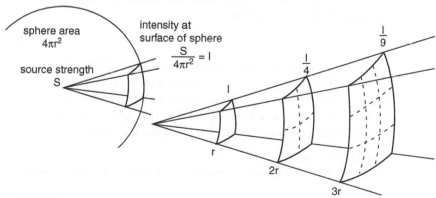

■ **FIGURE 2.8** Intensity from point sources. *From http://imagine.gsfc.nasa.gov/docs/science/try_l2/inverse_square.gif*

Suppose a security-related process was shown to be affected by a certain physical quantity. Furthermore, suppose the change in the amount of the quantity as a function of time is decreasing and is proportional to the physical quantity present. Let's say that a constant of proportionality that governs this rate of change has also been identified. We could write a simple expression that describes this model where C is the physical quantity, R is the proportionality constant determining rate, and t is time:

$$dC/dt = -RC \tag{2.3}$$

This expression is read as the rate of change with time of the quantity C (dC/dt, as recalled from your calculus days) equals the negative of R times the quantity C. The nature of R will surely vary according to the specific process but will always be expressed in units of "something per unit time" for exclusively time-dependent processes. Conveniently, there is a simple solution to this so-called first order differential equation where the time dependent solution is given by

$$C = C_0 e^{-Rt} \tag{2.4}$$

You no doubt recognize e^{-Rt} as the decreasing exponential function shown in Figure 2.7. We will see various forms of this function used repeatedly throughout this book since it describes many naturally occurring processes related to security.

As an example, let's say the quantity C represents the concentration of clean air in a room. The room air is assumed to be 100% clean before the outside contaminant is released. R is the rate at which the clean internal air is exchanged with the contaminated outside air and equals two room air exchanges per hour. C_0 is the initial concentration of clean air in the room (i.e., 100%). The product of R and t is a unit-less number and can be retrieved by simply taking the natural logarithm of each side of Equation 2.4, which yields

$$\ln(C/C_0) = -Rt \tag{2.5}$$

We might want to know when half the internal air is comprised of external air and therefore presumed to be contaminated. This is equivalent to knowing when the ratio of the concentration of contaminated air to the initial concentration in a room equals 50% (i.e., $C/C_0 = \frac{1}{2}$). Substituting the values into the above expression yields the following:

$$\begin{aligned}
\ln(1/2) &= -0.69 \\
&= -(R = 2 \text{ room air exchanges per hour}) \times (t = \text{time in hours})
\end{aligned} \tag{2.6}$$

Solving this equation for t we see that half the room is contaminated in approximately 0.35 hours or 21 minutes if the rate constant is two room air exchanges per hour. I will perform the exact same calculation, but this serves as a good introduction to security-related processes displaying exponential behavior. The rate constant, R, accounts for the effect of all processes that govern the exchange of bad air with good air and has a profound effect on risk in this case (see Section 5.4.3).

Although R is the proportionality constant determining how the quantity C changes with time, it is important to recognize that similar looking equations can govern very different processes. For example, this type of equation can be applied to quantities that change with distance. So I could just as easily write the following equation

$$dC/dx = -RC \qquad (2.7)$$

From the form of this expression it is clear that this looks identical to the time-dependent version except that the rate of change of the quantity C is expressed in terms of the scenario-dependent parameter distance, x. Implicit in this expression is the fact that R is given in units of "something per unit distance." In exact analogy with the time-dependent case, the solution is given by

$$C = C_0 e^{-Rx} \qquad (2.8)$$

This tells us that the quantity C decreases exponentially from its original value C_0 and at a rate governed by R. A graph of this function looks like the one shown in Figure 2.7.

For example, we might want to know the intensity of a beam of gamma radiation as a function of distance in a material through which it propagates to design an effective mitigation strategy. The curve would take the form of a decreasing exponential plotting beam intensity versus material thickness. The steepness of the curve would be determined by a different rate constant, which in this case is the mass absorption coefficient of the material. We will see more examples of physical processes characterized by expressions of this type in Part 2 of this book.

2.8 **VISUALIZING SECURITY RISK**

Succinctly and precisely capturing information related to security scenarios and incidents is important in revealing physical phenomena associated with the vulnerability component of risk and also in displaying trends. Understanding trends is extremely important in developing security strategies because these can reveal whether a security scenario is improving or worsening and at what rate. In addition, the manner in which data are portrayed is of great value in highlighting risk relevant issues and in focusing attention on those areas deserving the most attention.

Displaying the scale of physical quantities is best achieved with line graphs such as those for linear and nonlinear functions seen in Figures 2.3 and 2.4. A line immediately focuses the eye on the direction (i.e., increasing or decreasing) and the rate of change. If parameter values extend over a wide range, then logarithmic rather than linear values can be used for one or both axes. As observed in the semi-logarithmic plot of Figure 2.5, this had the effect of compressing the range of numerical values; therefore it is a convenient method of graphically displaying mathematical functions.

It is important to remember that unit increases in the value of the logarithm correspond to powers of 10 increments on a linear scale. So if log x is plotted versus x, a change from 1 to 2 in log x implies a change in absolute value of 10 to 100 (i.e., from 10^1 to 10^2). The Richter scale that measures the severity of earthquakes is expressed in logarithmic units. This is why a magnitude 6 earthquake is 100 times more severe than a magnitude 4. A logarithmic scale is merely a compressed version of a linear scale, which is precisely the point since it facilitates the display of functions that vary over a wide range of values. Another example of a semi-logarithmic plot is shown in Figure 2.9 to illustrate the point, and this

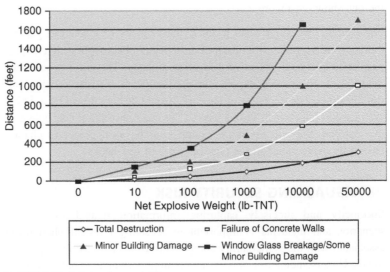

■ **FIGURE 2.9** Explosive effects on buildings.

graphic is shown again and appropriately referenced in Chapter 5. Here the net explosive weight is plotted in increments of powers of 10 (i.e., logarithmic) against the linear scale of distance.

I've noted that phenomena or scenarios that change with time are often best displayed with a line graph. A line highlights the ups and downs of cycles and the rate of change. Figure 2.10 again illustrates this point. This is a plot of room air contamination as a function of time for leaky and airtight buildings. I have also appropriated this particular graph from Chapter 5.

It is immediately clear from the shape of the plot that airtight buildings are infiltrated by outside air much more slowly than leaky buildings, which intuitively makes a lot of sense. (It is worth mentioning that it is possible to do curve fitting with linear plots in the popular Microsoft Office application, Excel.) This yields information on just how well the expression of interest "fits" standard functions such as exponentials, straight lines, quadratics, etc.

Pie charts and bar/column graphs can also be useful in displaying security information. Pie charts in particular are excellent at focusing attention on the fractional components of a program or budget. The pieces of the pie can have their values displayed on the chart as an absolute value or as a percentage of the total pie. Bar and/or column graphs are effective in displaying the absolute number of security incidents or their distribution among multiple venues. Some examples of simple security-related graphs are shown in Figures 2.11–2.14 and were created quite easily using the graphing wizard of Microsoft Excel.

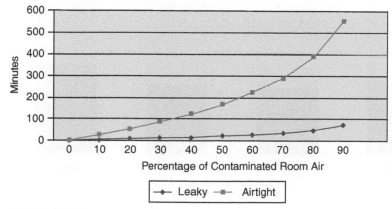

■ **FIGURE 2.10** Room contamination by outside air.

FIGURE 2.11 Thefts by building.

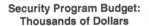
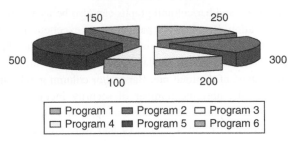

FIGURE 2.12 Security programs and budgets.

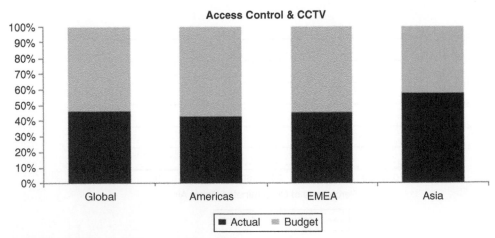

FIGURE 2.13 Global spending for access control and CCTV.

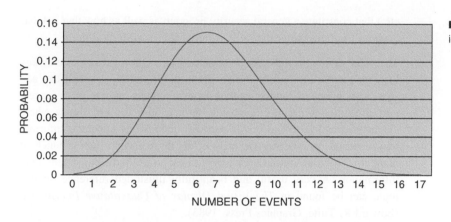

Another potentially useful display of security risk information is the so-called spidergram. It allows one to display multiple mitigation methods, their status or magnitude, and even their time evolution in one graphic. In the example shown in Figure 2.15, the points of the spider web represent the individual risk areas (e.g., physical security, information security,

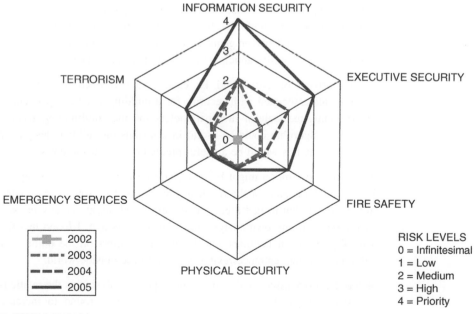

■ **FIGURE 2.15** Spidergram.

etc.) that constitute a general security program as well as how that risk has evolved from year to year.

The value of simple and well-constructed graphical images to convey key ideas or themes in security is difficult to overestimate. A graph can add significant punch to arguments on risk or even add weight to pleas for resources. A plot or graph immediately reveals structure and trends where characteristics like linearity will become immediately apparent. Information displayed in this way typically offers more insight than a bunch of numbers. There are indeed many ways to successfully display quantitative information. A very useful and beautifully illustrated reference on this topic can be found in *The Visual Display of Quantitative Information* (Edward R. Tufte, Graphics Press, 1983).

2.9 **AN EXAMPLE: GUARDING COSTS**

My guess is there are still some readers who are scratching their heads and wondering what any of this has to do with security risk. A simple but relevant example may help convince those of little faith.

The use of guards is typically quite important in a facility or campus security strategy and is arguably one of the more visible elements of that strategy. However, the deployment of guards can be relatively expensive since they often operate $24 \times 7/365$, earn overtime wages, and require training, health insurance, and uniforms. It would therefore be useful to develop a simple relationship between guard costs and significant drivers of building expenses to more easily anticipate costs and predict variances.

If such drivers of cost could be easily parameterized, the financial requirements associated with security for a building would in part entail understanding the scale of such parameters and then multiplying accordingly. This algorithm could be useful in decisions on building designs or acquisitions as this would be a useful predictor of recurring costs.

One initial assumption might be that guard costs mostly relate to square footage, building size, or the resident population. This sounds reasonable on the surface, but in fact a budget so formulated might actually be way off the mark. In my own analysis done some years ago, I found that the cost of guarding in facilities actually depended most on the number of active portals (i.e., entrances/exits) opening to the outside world.

Moreover, if one discovered through this type of analysis that the relationship between guarding costs and open portals was found to increase nonlinearly for some facilities (e.g., as the square or cube of the number of active portals), this would be a useful relationship for resourcing and overall risk management purposes. In the interest of full disclosure, a consistent scaling relation for this problem continues to elude me, but the goal is a noble one and is worth pursuing.

Such a relationship might imply that a small one-story building with five operational entrances/exits would actually be more costly to operate than a fifty-story building with two such portals, independent of building population. A logical explanation is that the major physical risk to the building is often found at the transition between the external world areas and the building interior, and these transition points require additional security controls.

Understanding this precise relationship might help campus designers who have the foresight to ask security managers for input prior to building construction. Most important, determining the key security parameters and understanding how changes in those parameters drive cost, risk, etc., are central to a risk management program.

2.10 **SUMMARY**

Understanding how physical quantities that affect security risk such as intensity or concentration change or scale with scenario-dependent parameters like time and distance is important in determining the vulnerability component of risk. Certain mathematical functions and units of measurement are frequently used to represent these quantities and therefore directly relate to security risk assessments. The concepts of linearity and nonlinearity are central to understanding the behavior of security-related quantities and in developing appropriate mitigation strategies. Graphical representations of the behavior of these quantities and program-related factors can yield valuable and even immediate insight into risk and associated requirements for mitigation.

Security risk measurements and security programs

Risk ... risk is our business. That's what this starship is all about.
That's why we're aboard her. — Captain James T. Kirk
"Return to Tomorrow," *Star Trek*, **Stardate 4768.3**

3.1 **INTRODUCTION**

The breadth of threats in the modern world of security can seem daunting. On a high level, these range from the relatively mundane like "theft" to the more esoteric variety such as chemical, biological, and radiological weapons. If one considers all the variations of each threat, in addition to differences associated with venue, facility type, etc., there are potentially an infinite number of threat scenarios. What are the implications of this to a modern security program, and more specifically, how do security directors ensure they are addressing each one?

Most security departments do not have infinite resources so either available resources must be rationed, threats prioritized, or both. From a programmatic perspective, the problem of a multitude of threat scenarios is compounded by a limited ability to test the efficacy of mitigation. The result is a potential perfect storm of uncertainty.

To be fair, most security programs will not be hamstrung in this way. But in general, a lack of rigor and/or process in assessing risk can contribute to an uncertainty in the effectiveness and efficiency with which security mitigation is apportioned and delivered.

The existence of multiple threat scenarios does not necessarily mandate an equal number of mitigation methods. Threats can and should be grouped together and more efficiently addressed in parallel. An example might be the threats of theft and physical assault, where the use of guards could simultaneously mitigate the risk associated with both threats. So sometimes the mitigation methods naturally, but not necessarily purposefully or strategically, drive the security program to be efficient in addressing the spectrum of threat scenarios.

With respect to effectiveness, in my experience few security programs rigorously validate or measure the performance of mitigation in reducing risk. This is not to say that these security programs are not effective, but there can be a lack of rigor in assessing and providing assurances of same.

Since threats are often dynamic, assessing the risk associated with a threat should be an ongoing process. Effective security risk mitigation consists of following a process of continuously monitoring, assessing, and adjusting mitigation based on current risk profiles. The elements of that process represent the crux of the material in this chapter. Ideally, the frequency of assessment will be greater than the rate of change of conditions that manifestly affect this risk profile.

I suspect that the information presented herein will be of most interest and/ or relevance to security professionals. This is because it specifies key elements of a standardized approach to security risk management and addresses a nearly universal problem facing security professionals. Due to their inherently defensive posture, security programs sometimes lack a strategic approach to their core business. Strategies tend to evolve more naturally in revenue generating entities where the bottom line automatically leads to an explicit performance metric. Therefore, I believe a focus on rigor and standardization will be embraced by security professionals, not only to improve effectiveness and efficiency but also to exploit analytic methods and the development of meaningful security metrics.

Given the potential complexity and scope of even a modest security program in today's environment coupled with the emphasis on cost containment, it seems that what is required is an overarching risk assessment process that facilitates effectiveness and efficiency *in measurable ways*. This can be accomplished by (1) identifying the unique threats of concern, (2) specifying the factors that enhance the individual components of risk for each unique threat, (3) establishing mitigation to address identified risk factors, and (4) invoking analytic tools and quantitative methods to establish metrics and measure the gap between threat and mitigation.

This chapter discusses the details of this process as depicted in Figure 3.1. This useful graphic shows threats as the progenitors of risk, the individual components of risk, and the continuous cycle of risk assessment, mitigation, and measurement.

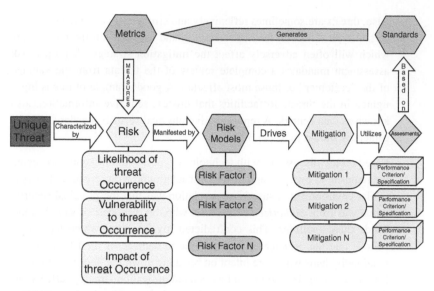

■ **FIGURE 3.1** The security risk assessment process.

3.2 **THE SECURITY RISK ASSESSMENT PROCESS**

3.2.1 **Unique threats**

It bears repeating that all security risk issues derive from threats. Moreover, it is a useful exercise to return to first principles when examining any security problem and to thoroughly understand each unique threat of concern. This increases the likelihood that mitigation will be correctly and efficiently applied. So in analyzing any security problem, the very first step is to accurately identify and categorize the set of unique threats.

The notion of uniqueness deserves further discussion since the risk assessment process is rooted in this idea. It is easy to be glib and use general terms like "theft," "terrorism," "information loss," etc., and plan mitigation on a high level based solely on this casual approach to categorization. However, differences in threats within these general categories can profoundly affect mitigation strategies.

For instance, all terrorists are not created equal. Attacks can be implemented in different ways and for different reasons that will significantly affect the respective likelihood and/or vulnerability components of risk. Some notable examples include the attack using vehicle-borne explosives perpetrated by Timothy McVeigh versus the lone gunman attacks by Charles Whitman (University of Texas) and Sung-Hui Cho (Virginia Tech). These two types of threats have little in common except to be generally considered terrorism.

Also, threats are sometimes reflexively and erroneously assumed to be of a certain type. This can lead to "swinging and missing" on the risk factors which will often adversely affect the mitigation strategy. A proper risk assessment mandates a complete review of the threats from the vantage of the "customer" or those most affected. A good example of this is highlighted in the threats to facilities that process sensitive information, also known as data centers. A representative physical security standard for data centers is shown in Table 3.5.

Based on my discussions with technology types, who unlike me possessed the requisite knowledge and historical background with respect to this particular environment, a surprising revelation was that the principal security threat was from *authorized* individuals who did dishonest or stupid things inside these facilities. This contradicted my intuition, which led me to believe that the principal threat was due to damage or theft by *unauthorized* individuals who were intent on breaking down the door to gain physical access. The risk factors and associated mitigation strategy reflects this distinction and the recognition that these represent unique threats that potentially require dissimilar mitigation.

3.2.2 **Motivating security risk mitigation: The five commandments of corporate security**

The set of unique threats ultimately drives the requirements for security risk mitigation. And each risk factor associated with a unique threat is the *raison d'être* for a given form of mitigation. At the highest level, however, there is a set of underlying requirements that motivates any mitigation strategy.

For example, it may seem obvious that the motivation to use guards is to establish a safe and orderly environment. This requirement, along with four others of a similar nature, forms the philosophic underpinnings of security mitigation at a facility. Collectively and somewhat grandiosely, I choose to call these "The Five Commandments of Corporate Security." All mitigation methods should ultimately link to one or more of these requirements. Moreover, these represent the most general requirements for a security risk mitigation strategy. Finally, and so as not to appear to be in competition with the more venerated principles of a similar name, no claims are made regarding any theistic origins of these commandments.

1. Only properly authorized and authenticated individuals should be afforded physical access to company-occupied or -controlled space.
2. The internal and immediate external environments of company-occupied or company-controlled space should be safe and secure at all times.

3. Company-sensitive and/or confidential/private information should only be made accessible to properly authorized individuals who have a legitimate "need-to-know."
4. Company personnel should be expeditiously made aware of potentially unsafe and/or business-threatening conditions along with guidance on risk mitigation whenever possible.
5. At a minimum, company-occupied or -controlled space must be in compliance with all applicable fire and health/safety statutory codes and requirements.

3.2.3 **Security risk models**

Once the set of unique threats is identified, the next task in assessing the risk associated with a unique threat is to identify the risk factors as noted in Section 1.3. These enhance the likelihood and/or vulnerability components of risk for each unique threat. These factors constitute the risk model. Identifying the risk factors is an enormously important exercise, since these should drive the requirement for each and every mitigation measure.

Suppose you are the security director of a large, Manhattan-based company and the big shots are thinking of either purchasing or building a new facility. They presciently solicit advice from the director of security and wisely intend to factor your recommendation into their decision before a financial commitment is made. How would you proceed and how would you present the security-related information to maximize its effectiveness for the intended audience?

The first point of business is to identify the set of unique threats. These might include industrial espionage, terrorism, and street crime but not necessarily in that order of priority. Some justification of your choice of threats would probably be welcome, possibly with some supporting statistics and/or information from reliable sources. The next phase of your presentation might entail identifying those factors that affect the magnitude of the likelihood and vulnerability components of risk associated with those unique threats.

Regrettable, company management is already fixated on a particular site and is looking to the security department for a recommendation with regard to that venue. You first focus on the potential for threat occurrence for significant threats which might impact your company.

One of these threats is terrorism which can come in all shapes and sizes. A significant security concern in urban America is the use of vehicle-borne explosives by fundamentalist groups with anti-Western sentiments. Of course, we only have to look at the bombing of the Alfred R. Murrah building in Oklahoma City to see the same method used to further a completely different cause. Although the terror *modus operandi* for anti-Western groups and anti-government separatists happens to be identical, the philosophic basis upon which they operate is very different and could influence the potential for a terrorist incident at a particular facility. Therefore, you proceed with caution and precision in analyzing the risk factors associated with terrorism relative to your company's facility.

Suppose your company, USA Potato Chips, manufactures potato chips consistent with its name. It might be logical to assume that USA Potato Chips would not be *numero uno* on any terrorist hit list except perhaps for those groups that feel passionately about protecting the rights of potatoes.

However, if you are *the* world leader in manufacturing potato chips, and have also been identified as a prominent American company, the potential for threat occurrence might not be so negligible. As the security director of USA Potato Chips, there is probably little you can do to influence the potential for a threat to occur against your facility. The risk factors affecting this component of risk are inevitably beyond your control. But you could possibly do a lot to decrease the vulnerability component of risk. For example, you might recommend installation of a perimeter barrier as a control to ensure sufficient vehicle standoff. Vehicle proximity is a significant risk factor with respect to the vulnerability component of risk for the threat of vehicle-borne explosives. Identifying those factors that affect the individual components of risk is the essence of a risk model.

If there is any "good news" in this instance it is that the potential for an attack against your facility should not vary much from site to site; that is, if your company is considered a likely target because of brand name recognition and a linkage to the United States, then the potential for incident occurrence should be roughly equivalent irrespective of location, at least within the same city. This might not appear to offer much consolation, but silver linings can be frustratingly elusive in the security risk business.

Now consider a very different but related scenario; namely, the potential for collateral damage resulting from an attack by vehicle-borne explosives against facilities located in proximity to USA Potato Chips. How would that change the risk model? Here the risk factors affecting the potential for a security incident would be profoundly location-dependent. Suppose you changed jobs and are now the security director of Joe's Hardware, a company that is searching for a new location for its flagship store. Based on your risk assessment, it is clear to you that a prominent American potato chip manufacturer represents an attractive target to organizations intent on blowing up iconic American symbols.

In this case you would probably be disinclined to select a site that was anywhere near USA Potato Chips unless you had no viable alternative (i.e., you are being pressured by senior management). Therefore, the real estate broker assisting Joe's Hardware in its search for a new venue has been told to avoid showing sites anywhere close to USA Potato Chips if she hopes to earn a commission on the deal. The security director at Joe's Hardware has just successfully implemented risk mitigation in accordance with the risk model.

Conclusions about the potential for threat occurrence in this instance are inherently subjective, but nonetheless real, with actionable alternatives for risk mitigation. Although there is no way to meaningfully quantify the likelihood of a future vehicle-borne explosive incident, it has been determined that the risk factor of physical proximity to USA Potato Chips definitely enhances the potential for incident occurrence. Notice there was no mention of any (highly speculative) odds of an attack against a particular facility at a specific time.

Situations like these pose ongoing challenges for security professionals. It is important to avoid the temptation to provide a number for the likelihood component of risk quantifiable. Often this is done to enhance the perception of accuracy when in fact this can have the exact opposite effect if the details are subject to scrutiny. Instead, one must leverage the available source information and apply good judgment to arrive at a qualitative but nonetheless reasoned opinion on risk.

Recall that the risk assessment of USA Potato Chips focused on the risk factors associated with a particular version of the threat of terrorism. A similar exercise is required for each such unique threat including dissimilar varieties of terrorism, if that is relevant.

So if Joe's Hardware believed banks were a likely target of violence with a knock-on effect of business disruption or violence affecting its employees, then facilities near banks would be off-limits as well with respect to site selection. Continuing with its risk assessment, Joe's Hardware might also be concerned with the threat of street crime. Which risk factors affect the potential for occurrence for this unique threat? Certainly locations near high crime areas might be a risk factor where part of the assessment might involve flushing out the 24/7 characteristics of the neighborhood.

But Joe's Hardware is not finished in its evaluation of the threat from vehicle-borne explosive attacks. Evaluating the vulnerability component of risk is equally important in this case. Remember the ichthyologist who compensated for extreme vulnerability by assuming a low potential for occurrence. Suppose Joe's Hardware has received an irresistible deal for a facility across the street from USA Potato Chips. If the senior management at Joe's Hardware chooses to pursue this option, the only viable recourse from a security perspective is to compensate for the enhanced potential for threat occurrence by decreasing the vulnerability component of risk.

As it happens, the savvy real estate broker understands the security concerns of Joe's Hardware. She has informed the security team that the selected facility has windows that are protected by anti-blast film. After reading this book and obviating the need to hire a high-priced physical security consultant, the security director analyzed the specifications on the window film relative to the effects of likely blast scenarios. The security director was thus able to provide an informed view of the vulnerability component of risk to senior management along with a reasoned recommendation on mitigation.

In addition to the decrease in vulnerability afforded by anti-blast window film, the security director of Joe's Hardware has also been able to quote some savings associated with choosing a facility with pre-installed mitigation, although to be fair these might have been factored into the price of the building. Clearly, such an analysis has yielded a more informed and nuanced assessment of risk and the security director will no doubt reap the benefits when generous year-end bonuses are distributed to the entire security team!

Note that risk factors for the impact component of risk were not included in the risk model. This is not to say that this component of risk is not relevant. In some sense it is *the* most important component of risk. From a pragmatic perspective, why develop a strategy for a threat that has no impact on your organization? For example, the director of security for a car dealership in Miami is rightfully very concerned with the trend in car thefts in South Florida but not particularly bothered about tornadoes in Topeka, Kansas.

Also, decisions regarding the impact component of risk tend to be organization-specific. If a threat is on a security department's radar screen then it must meet some internally determined impact threshold that could easily vary from company to company. What is impactful to General Motors in Michigan will not necessarily be significant to General Dynamics in Connecticut.

Mitigation to address the risk factors for "standard" threats such as terrorism, violence against employees, theft, etc., have been incorporated into simple guides to facility and site selections listed in Table 3.1. Specifically, they contain site and building-specific features intended to address factors that enhance the likelihood and vulnerability components of risk associated with some common threats to the facilities. Ideally these features should be considered before signing a lease or occupying a facility.

Table 3.1 Guidelines for Facility and Site Selections
Site-related security guidelines
1. Avoid the following:
a) Iconic, trophy, historic, listed, or high-profile sites and/or locations near such sites b) Uncontrolled public facilities for vehicles (e.g., tunnels, parking areas, etc.) directly beneath or adjacent to the site
2. Seek the following:
a) Maximum setback from the street on all facades b) Maximum physical separation from neighboring buildings c) Convenient external assembly points d) Close proximity to emergency services e) Easy access to major roads or arteries
Facility-related security guidelines
1. Seek the following in conjunction with a proper facility risk assessment before signing a lease:
a) Sole building occupancy or sole floor occupancy at a minimum b) Physical access-controlled building entrances and exits to include parking facilities c) Structural designs that minimize the risk of progressive collapse in the event of an explosive incident d) Buildings with appropriate blast mitigation measures e) Effective acoustic isolation for internal offices/conference rooms next to non-company-controlled space f) Provisions for proper visitor access and control g) Elevated and physically secured HVAC air intakes h) Fire detection/prevention and life safety systems that meet company standards as well as all applicable codes i) Adequate emergency escape routes j) Internal space with the potential for segregated mail sorting/distribution k) Appropriate access controls for on-site parking and preferably not located beneath the building if a multi-tenant facility l) Provisions for secure equipment storage

3.3 **MANAGING SECURITY RISK**

3.3.1 **The security risk mitigation process**

As stated in the beginning of this chapter, it is often helpful to think about a security problem from first principles. There should be logic to the risk assessment process, specifically in determining the mitigation required to address the risk factors for each unique threat. As always, the process originates with an understanding and categorization of unique threats and the associated risk factors that enhance the individual components of risk.

The sole purpose of mitigation is to minimize the likelihood or vulnerability components of risk. With respect to the threats of theft and violence against employees, the fact that many people steal and/or are violent is a real concern. It is a general fact that the past is prologue with respect to human behavior, so background checks are considered an effective means of mitigating risk factors associated with the likelihood component of risk for employees considered for hire.

Referencing the First Commandment of Corporate Security, the requirement to allow only properly authorized and authenticated people inside facilities is driven exclusively by the historically bad behavior of many members of our species. If all people were honest, nonviolent, and trustworthy, this commandment would not exist.

Similar logic applies to the other commandments. The reason near-biblical importance has been attached to these commandments is because of their general applicability and fundamental nature. They represent the basis for the drivers of all security risk mitigation. This may appear overly hyped but it is true. And unfortunately there is a tendency to reflexively apply risk mitigation without thinking precisely about the underlying goals. This can result in unnecessary, unproductive, and/or inefficient security measures.

The goal of a security professional as a risk manager is to identify and implement a set of effective mitigation methods for each unique threat, and this often calls for threat prioritization. Although each method may be intrinsically valid and hopefully even work, it says nothing about the effectiveness and efficiency of a mitigation *strategy*. In other words, how do we know whether the set of mitigation measures is applied optimally to priority threats and in proportion to the risk? We require a process wherein mitigation is applied coherently and consistently across the range of disparate threat scenarios.

I have already highlighted the importance of accurately identifying and categorizing the set of unique threats at the beginning of any assessment so I will proceed from there. In thinking about security solutions, indeed about most problems, it helps to start with broad but distinct generalities and work toward specifics. Many people do the reverse and this can have difficult consequences.

In fact, addressing security issues can ultimately boil down to a problem in classification. The goal is to put the problem into general categories or "food groups" of unique threats and then identify the risk factors for each category. So following the critical exercise of threat and associated risk factor categorization, we establish a set of high-level mitigation measures called "controls" that are necessary to address each risk factor in the risk model. These controls might include very general security functions such as "visual monitoring," "physical access control," "communication devices and protocols," etc. In this mitigation process, at least one control must be specified for each risk factor. The controls do not constitute a strategy by themselves, but they represent a significant step in the right direction.

Moving from the general to the specific, a method or set of methods should be identified and used to implement each control. The control of visual monitoring is illustrative in this instance. Visual monitoring describes a general function intended to increase an awareness of what is going on in the environment in real time as well as to facilitate forensic or post-incident analyses. Presumably this will assist security personnel to respond efficiently and effectively to incidents. But how is visual monitoring implemented? The only commercial methods of visual monitoring I am aware of include the use of guards and closed circuit television (CCTV).

For a unique threat such as general crime, a control of visual monitoring might be specified. CCTV and/or guards are then selected as the methods to implement that control. There could be other controls required for this threat, corresponding to each unique risk factor, but once specified, each control would require at least one method of implementation. I note in passing that the underlying motivation for invoking mitigation to address the threats of general crime and/or terrorism is derived from Commandment Two of the Five Commandments of Corporate Security. This states that it is necessary to establish and maintain safe and secure environments. Amen.

It is all well and good to specify that a control in the form of visual monitoring is required to address a risk factor and even to identify a method to implement that control. However, the specific method chosen must perform at a certain level to be effective. For example, and sticking with CCTV as an implementation method, it could be important to be able to identify an individual from either a recorded or live image. This performance criterion is sometimes referred to as facial recognition.[*]

Note that facial recognition is a very different standard of performance for visual monitoring than so-called "situational awareness." Whereas facial recognition requires the operator to be able to recognize Carl Young from his image as it appears on a monitor, the latter criterion refers to the need for a visual overview of an area to understand what is going on in a general area.

Indeed, technical specifications such as the lens depth of focus, frame rates for recording, lighting, etc., in the case of CCTV are likely to be quite different for each distinct performance criterion. It goes without saying that the chosen method and associated performance criterion should be driven by the stated operational requirement for installing the camera in the first place. Security technology staff should be insistent in receiving unambiguous operational requirements, preferably in writing, prior to installing all security equipment. This is an essential element of any security program that is hoping to avoid the pitfalls of "security theater".

Another example helpful in illustrating the risk mitigation process is the decision to use bollards to provide a secure perimeter (see Section 6.4). Let's say that a unique threat has been identified in the form of vehicle-borne explosives. A risk factor associated with the vulnerability component of risk has been further identified as physical proximity by unauthorized vehicles. Therefore, some type of barrier has been specified as a necessary control to address this risk factor.

In progressing from the general control of barriers to the specific method needed to implement that control, bollards have been selected to ensure a secure perimeter. An appropriate performance criterion for these bollards has been formulated so that a vehicle must not be able to penetrate a distance greater than one meter following impact. The technical specifications for the bollard to ensure that the performance criterion is satisfied will be based on an analysis of the vulnerability to a vehicle being able to penetrate the bollard line.

---

[*]Facial recognition must be distinguished from facial identification. The former refers to the ability of an individual to recognize a known individual from a CCTV image. The latter represents the ability of a system to identify an individual based on a comparison of a CCTV facial image with facial images stored in memory. Facial identification is considered a much more difficult technical problem.

Specifically, if threats in the form of van-sized vehicles are of most concern, the maximum run-up velocity and estimated weight of a typical van is used to calculate its kinetic energy on impact at the required standoff distance. This should be compared to the proposed bollard technical specifications. For example, it wouldn't do much good if the bollards of choice were constructed from rubber to stop a speeding van intent on breaching a secure perimeter.

In general, the methods chosen to implement each control must have specific performance criteria and associated technical specifications for the control to be considered effective in addressing the identified risk factor(s). Performance criteria and associated technical specifications for four common methods, CCTV, physical access systems, exterior barriers, and window blast protection are listed in Appendices D to G, respectively.

In summary, the process used to develop a particular security mitigation strategy is straightforward: (1) identify the set of unique threats, (2) determine the relevant risk factors that enhance the likelihood or vulnerability components of risk associated with each unique threat, (3) establish the full set of controls to address each risk factor, (4) specify the methods necessary to implement each control, and (5) ensure that each method achieves the required performance criteria according to stated operational requirements by adhering to technical specifications that are in compliance with agreed security standards.

The security risk mitigation process is captured in Figure 3.2.

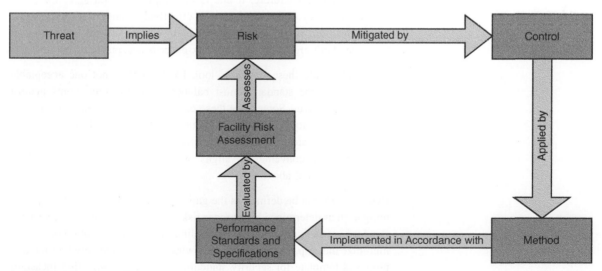

■ **FIGURE 3.2** The security risk mitigation process. *Graphic developed by Caroline Colasacco.*

3.3.2 **Security risk standards**

The risk mitigation process specified in this chapter offers a convenient segue to a template for a risk-based, security standard. Standards are an essential part of security mitigation as well as the overall risk assessment process. Why?

One reason is that the use of security standards might be the only available mechanism to rigorously evaluate risk and establish a meaningful security risk metric. In a defensive world such as security this is critically important. How do you know that a risk mitigation strategy is effective in managing risk? Is the fact that a facility or organization has not suffered an incident indicative of a satisfactory security strategy or does it signify indifference on the part of an adversary? Ultimately, an adversary may decide to conduct their own "test" and by then it might be too late. Unfortunately it is often difficult to self-test or "red team" security mitigation methods with statistically significant confidence, although this is always a fantastic idea whenever possible and practical. Therefore, indirect methods must often be used to validate the sanity and integrity of security strategies.

The logic behind the general use of physical security standards as part of the risk assessment process is simple but irrefutable; namely standards represent the minimum requirements necessary to effectively manage the risk associated with the set of unique threats. More specifically, they specify the controls and methods that are deemed necessary to reduce the likelihood and vulnerability components of risk to a manageable level. A combination of judgment and rigorous analyses should be used to develop these standards. Therefore, if one is in compliance with agreed security standards then by definition the risk is effectively managed. Conversely, if an organization is not in compliance with agreed security standards, or those standards are out of date, the risk exceeds acceptable levels.

But what should these standards look like? There is not one acceptable structure, but the standards must balance security requirements against business objectives. Sometimes these can be at odds with each other. This calls for flexibility but not to the point where standards lose their effectiveness. Fortunately, there is a structure that allows for both rigor and flexibility and conveniently aligns with the risk mitigation process described in detail above.

Residual risk can be defined as the gap between the threat and the applied mitigation as determined by a proper risk assessment. An effective security standard is one that provides a formal link to the risk mitigation process to minimize that gap. Therefore, the following should sound very familiar as a proposed template for security standards: high level mitigation measures called controls are specified for each risk factor associated with each unique threat. All controls are deemed essential to effectively manage the risk associated with the identified set of unique threats. A method or set of methods is required to implement each of these controls.

The number and type of methods that apply in each instance is a judgment call and results from measurements, calculations, and/or estimations of vulnerability and likelihood as discussed extensively in Part 2 of this book. Such decisions should also be influenced by intuition gleaned from professional experience. Finally, performance criteria with associated technical specifications ensures that identified security methods function according to specification under operational conditions. In other words, no rubber bollards please.

The key point is that the proposed template for standards incorporates the risk mitigation process into its very structure. Therefore, if a facility meets the defined security standard(s) then by definition the residual risk is effectively managed. Security standards so designed will help ensure that mitigation is applied in a consistent manner across similar risk profiles, an especially important goal for complex and/or global security programs. They will also offer both the rigor and flexibility required to assess and mitigate disparate threats in the face of competing priorities.

As always examples help to provide clarity, so I include a few here to illustrate the point. The following standards are not meant to be exhaustive and are not necessarily complete. Companies should endeavor to develop a full set of standards to address the threats and other relevant security issues appropriate to their respective environments.

One candidate security standard applied to office facilities is shown in Table 3.2. It is organized by area within a generic office building, and is replete with a full complement of controls and methods to address the risk factors associated with garden variety security threats such as theft, terrorism, and information loss. However, in customizing such a standard for a specific facility it is important to specify the threats and risk factors explicitly to ensure a complete and accurate mapping of risk factors to controls.

Another standard that sometimes confounds security and safety professionals is fire safety. A proposed standard using this structure is shown in Table 3.3. Because jurisdictions have specific code and associated requirements, a general standard has often eluded corporate managers looking for a common framework. Since this book generally avoids issues associated with fire or life safety, this standard is provided only to showcase how the general approach is applicable to disparate and/or "difficult" programs.

Table 3.2 Physical Security Standard for Office Facilities

	Method 1	Method 2	Method 3	Method 4	Method 5	Method 6	Method 7
Perimeter and External Environs							
Control 1 = Visual Monitoring	CCTV (note: *no IP-based technology in non-controlled or exterior space*)	Guard Monitoring and Response Protocols					
Control 2 = Physical Control	Guard Monitoring and Response Protocols	Explosive Detection Patrols (canines and/or detection technology)	Vehicle control	Door Lock with Key Management	Paid Police Detail	Nighttime Lighting	Controlled Pedestrian Flow
Control 3 = Physical Stand-off from Public Areas	Barriers/Bollards/ Road Blockers	Distance or Set-back from Street	Paid Police Detail				
Control 4 = Risk Information Monitoring	Open Source Research/ Monitoring/ Alerting	Liaison/Law Enforcement Contacts					
Control 5 = Emergency Preparedness	Business Continuity Plans	Drills and Preparedness Exercises					
Control 6 = Exterior Facility Hardening	Explosive Window Treatment/ Mitigation	Ballistic Window Treatment/ Mitigation	Enhanced Structural Design of Façade and Internal Components	Distance from Vehicular Traffic	Distance from Iconic Structures		
Lobby/Reception Area							
Control 1 = Visual Monitoring (situational awareness for general floor area, facial identification at turnstiles and	CCTV	Guard Monitoring and Response Protocols					

access/egress points					
Control 2 = Authentication/ Authorization	Authentication via ID Picture/ Guard Verification	Authentication via Visual Recognition (applicable to smaller offices)	Authentication and Authorization via Two-Factor/ Biometric	Authentication and Authorization via Two-Factor/ID Card Reader and Pin Pad	Authorization via Global Access Control/ID Card-Triggered Access
Control 3 = Client/Visitor Management	Reception Services	Guard Monitoring and Response Protocols	Signage		
Control 4 = Physical Access Control	Turnstile/ Movable Barrier	Guard Monitoring and Response Protocols	Reception Services	Elevator Control	ID card Monitoring (e.g., lost, stolen cards)	Door with Locking Mechanism
Control 5 = Access by Emergency Services	Lock Override	Availability of Access ID Card	Guard Monitoring and Response Protocols		

General Office Space

Control 1 = Visual Monitoring	CCTV (situational awareness)	Guard Monitoring and Response Protocols			
Control 2 = Storage of Valuables	Safe with Combination/ Key Control	Locked Receptacle with Key Control			
Control 3 = Disposal/ Destruction of Documents	Cross-Cut Shredders	Appropriately Labeled Receptacles	Approved Waste Disposal Vendor		
Control 5 = Authentication/ Authorization	Authentication via ID Picture/ Guard Verification	Authentication/ Authentication via Visual Recognition (applicable to small offices)	Authentication and Authorization via Two-Factor/ Biometric	Authentication and Authorization via Two-Factor/ID Card Reader and Pin Pad	Authentication via ID Card-Triggered Access
Control 6 = Physical Access Control	Global Access Control/Id Card-Triggered Access	Alarm Interfaced to Security			

(Continued)

Table 3.2 Physical Security Standard for Office Facilities—*Cont'd*

	Method 1	Method 2	Method 3	Method 4	Method 5	Method 6	Method 7
Loading Dock							
Control 1 = Contractor/ Vendor Authorization/ Authentication	Electronically Pre-Authorized Contractors/ Vendors	Paper List for Pre-Authorized Contractors/ Vendors					
Control 2 = Vehicle Screening	Explosive Detection (randomly selected vehicles)	Explosive Detection (all vehicles)	Visual Inspection Based on Pre-Authorization				
Control 3 = Driver Authorization/ Authentication	Presentation of Drivers License/ Government-issued ID Compared to Contractor/ Vendor Access List	Visual Recognition Based on Contractor/Vendor Familiarity					
Mail/Package Processing Area							
Control 1 = Visual Monitoring	CCTV	Guard Monitoring and Response Protocols					
Control 2 = Physical Access Control	Global Access Control/ID Card-Triggered Access	Guard Monitoring and Response Protocols					
Control 3 = Secure Storage	Safe	Locked Receptacle with Key Control					
Control 4 = Authentication/ Authorization	Authentication via ID Picture/ Guard Verification	Authentication via Visual Recognition (applicable to smaller offices)	Authentication and Authorization via two-Factor/ Biometric	Authentication and Authorization via Two-Factor/ID Card Reader and Pin Pad	Authorization via Global Access Control/ID Card-Triggered Access		

Child Care Area					
Control 1 = Visual Monitoring	CCTV	Guard Monitoring and Response Protocols			
Control 2 = Physical Access Control	ID Card-Triggered Access	Door Lock with Key Management	Guard Monitoring and Response Protocols		
Control 3 = Emergency Alerting	Alarm/Duress System Interfaced to Security	Intercom interfaced to Security	Guard Monitoring and Response Protocols		
Technology Room					
Control 1 = Visual Monitoring	CCTV				
Control 2 = Physical Access Control	ID Card-Triggered Access	Anti-piggybacking	Door Lock with Key Management	Alarm Interfaced to SCC/Security	
Control 3 = Authentication/Authorization	Authentication via ID Picture/Guard Verification	Authentication via Visual Recognition (applicable to smaller offices)	Authentication and Authorization via Two-Factor/Biometric	Authentication and Authorization via Two-Factor/ID Card Reader and Pin Pad	Authorization via ID Card-Triggered Access
HVAC Area					
Control 1 = Physical Access Control	Door Lock with Key Management	ID Card-Triggered Access	Guard Monitoring and Response Protocols	"Door Detective" Interfaced to CCTV and/or SCC/Security	Restricted Physical Access to External Vents (elevated vents preferred or guard monitoring response protocols)
Control 2 = Visual Monitoring	CCTV	Guard Monitoring and Response Protocols	Alarm/incident Alerting Interfaced to Security/Scc		

(Continued)

Table 3.2 Physical Security Standard for Office Facilities—*Cont'd*

	Method 1	Method 2	Method 3	Method 4	Method 5	Method 6	Method 7
Stairs and Emergency Egress Point							
Control 1 = Compliance with Local Fire Code	Fire Department Inspection and Certification						
Control 2 = Physical Access Control	Door Lock to Restrict Inter-floor Access	Guard Monitoring and Response Protocols	Visual Authorization (smaller facilities)				
Control 3 = Building Egress	Lighting	Signage	Public Address System	Guard Monitoring and Response Protocols			
Control 4 = Visual Monitoring	CCTV	Guard Monitoring and Response Protocols					
Medical Areas							
Control 1 = Secure Storage of Medicines	Safe	Locked Receptacle/ Cabinet With Key Control					
Control 2 = Physical Access Control	Reception	ID Card-Triggered Access	Door Lock with Key Management				
Control 3 = Visual Monitoring	CCTV	Guard Monitoring and Response Protocols					
Security Control Room							
Control 1 = Physical Access Control	Global Access Control/ID Card-Triggered Access	Guard Monitoring and Response Protocols					
Control 2 = Visual Monitoring	CCTV	Guard Monitoring and Response Protocols					

Basement/Roof (also see HVAC Controls/Methods)

Control 1 = Physical Access Control	Locked Entry with Key Control	Guard Monitoring and Response Protocols	"Door Detective" Interfaced to CCTV and/or Security	Global Access Control/ID Card Triggered Access	Alarm Interfaced to CCTV and/or Security
Control 2 = Visual Monitoring	CCTV	Guard Monitoring and Response Protocols			

Conference Room

Control 1 = Restricted Audibility	50 dB of Acoustic Isolation Between Inside and Outside (if conference room is contiguous with non-controlled space)	Guard Monitoring and Response Protocols During Meeting	Conference Room Contiguous with Controlled Space	
Control 2 = Physical Access Control	Locked Entry with Key Control	Guard Monitoring and Response Protocols Prior to Meetings	Global Access Control/ID Card-Triggered Access	
Control 3 = Physical Inspections	Visual	Technical Surveillance Countermeasures (TSCM)		

Elevator

Control 1 = Emergency Communication	Intercom Interfaced to Security
Control 2 = Compliance with Local Building Safety Code	Successful Safety Inspection

Parking Areas

Control 1 = Physical Access Control	Tenant Authorization via Guard	Tenant Authorization via Card Access System
Control 2 = Visual Monitoring	CCTV	Guard Monitoring and Response Protocols

Table 3.3 Standard for Fire Safety

Threat	Controls to Protect People and Property	Methods to Enable Controls (Implemented According to Risk)
Fire	Risk assessment, including building/occupant profiling	Building specific attributes, geographic location, and number/type of building users
		Anticipated likelihood of fire
		Anticipated severity and potential spread of fire
		Ability of the structure to resist spread of fire and smoke
		Consequence of danger to people in and around the building
	Means of detection and warning	Detection system (smoke, heat, VESDA, manual call point)
		Alarm system (bell, siren, public address, LED display)
		Means of automatically notifying the fire department upon activation of an alarm
	Means of escape leading to a place of safety away from the building	Fire exits (minimum of 2 exits, travel distance, exit width)
		1 hour fire-protected corridors and 2 hour fire-protected stairwells with re-entry
		Preferred 120 people through a 44-inch stairway, maximum 220 people through a 44-inch stairway
		Emergency egress lighting throughout the building and escape routes; specific signage (including photo luminescent markings); all exit and directional signs throughout the building connected to emergency batteries and/or emergency generators
		Disabled egress (refuge, communication panel, evacuation chair, evacuation lift)
	Evacuation strategy	Total (simultaneous) evacuation
		Phased evacuation
		Progressive horizontal evacuation
		Shelter-in-place
	Means of limiting fire spread within a facility, or between tenants in a multi-tenant facility	Fire compartmentalization (minimum 1 hour fire-rated walls, floors, ceilings, and surface linings)
		Fire suppression (gas, foam, or water)
		Smoke curtains, fire dampers, and fire shutters
		Stair lobby protection
		Fire-stopping and cavity barriers
		Internal finishes/furnishings resistant to flame spread over surface (flame-proofing affidavits)
	Firefighting access and provisions	External road access
		Internal fire stair and lift
		Fire extinguishers (water, foam, powder, or CO_2)
		External water hydrants and internal stair risers
		Fire panel and control room
	Emergency response	Staff fire drills
		Fire warden training and staff awareness
		Assembly point

Table 3.3 Standard for Fire Safety—*Cont'd*

Threat	Controls to Protect People and Property	Methods to Enable Controls (Implemented According to Risk)
	Fire safety management/operations	Fire safety management team
		Fire systems testing and maintenance
		Good housekeeping (free access to emergency exits, separate storage of combustibles and flammables, etc.)
		Preventative maintenance schemes for all passive and active fire safety systems
		Maintenance of a building/facility fire safety log
		Implementation and compliance with facility fire safety document
		Fire exit doors self-closing and re-latching
		Certified fire safety officers
		Response teams
	Means of limiting external fire spread	Boundary separation distance
		External façade design (size of glazed openings)
		Roof and façade material and construction
Smoke	Smoke control/ventilation	Stairwell pressurization
		Mechanical smoke control system (smoke extraction)
		Natural vents shafts
		External wall and ceiling openings
		Basement ventilation system
Vapor explosions	Ventilation	Mechanical and natural ventilation
		Pressure relief valves/vents
	Detection and suppression	Gas detection
		Foam suppression
	Special construction	Explosion-resistant construction
		Wall/dikes to prevent leakage or spilling of flammable liquids

Notice the threats in this case all reduce to common issues like "fire," "smoke," etc. The controls required to address the risk factors are equally universal such as "fire suppression," "fire/smoke detection," "evacuation," etc. It is the specific *methods* invoked to enable these controls that will vary based on the jurisdiction as will the performance criteria and associated technical specifications for those methods.

This standard is definitely *not* intended to replace local fire codes. However, it might be helpful in identifying big gaps in coverage. It might also inform a manager on diverse resource requirements, highlight inconsistencies, or at the very least provoke discussion. In general, having a common, global standard can help to facilitate consistent approaches to program management.

Another common security issue is the storage of sensitive documents or other physical media, especially off-site. Debate regarding the proper controls and methods is always encouraged and is even useful. But it is the structure of the standard and its general applicability that should be appreciated here. Recall that all controls are required but the number and type of methods implemented (a minimum of one method per control) is left to the discretion of the assessor. An example of such a standard is shown in Table 3.4.

Table 3.4 Security Standard for the Off-Site Storage of Media Containing Sensitive or Proprietary Information

Threat	Controls	Methods to Enable Controls
Loss or theft of physical documents or other media	Proper packing of files	Relevant businesses witnessing packing
		Container labels
		Sealed containers (various methods)
	Secure transport of files	Locked transport vehicle
		Approved route from office to storage facility where required
	Documented inventory, storage and retrieval activities	Written receipt obtained from vendor prior to transport
		Written confirmation of arrival at storage facility
		Distribution of written report post incident
	Background-checked vendors	Vendor check
	Access control to facility and/or storage area	Identification of vehicles prior to entering the loading/unloading area
		Electronic authentication (card reader, pin pad, etc.)
		Key management system
		Authorized access list management (security officer)
		Escorting of unauthorized personnel
	Visual monitoring of facility and/or storage area	Monitored CCTV
		Security officer patrol (24-hour)
Information loss/unauthorized viewing	Appropriately secured files	Sealed containers
		Restricted access to files (authorized or escorted personnel only)
Damage to, or destruction of, physical documents or other media	Proper storing and handling of files	Fire- and water-resistant containers and/or storage area
		Unpacking/salvage procedures conducted by authorized personnel only
	Fire protection measures consistent with the value and/or criticality of the stored materials (as determined by the owners)	Relevant fire certificate is up to date

Companies often process electronic information within special facilities that are commonly known as data centers. Hence, a security standard for data centers is both important and useful. It is interesting to highlight the threats specified in this particular standard as these might be counterintuitive. As always, threats should ultimately drive the mitigation strategy.

In particular, note the "threat of damage by authorized and authenticated individuals inside data fields." As first discussed in Section 3.1.1, concerns over illicit or stupid actions by individuals who have authorized access to high risk areas are of equal or greater concern than the threat from intruders. It is therefore important to distinguish these as unique threats. The potentially dissimilar controls and methods specified in Table 3.5 reflect the mitigation intended to address the risk factors associated with each unique threat.

Table 3.5 Security Standard for Facilities that Process Electronic Information (Data Centers)

Threat	Controls	Methods to Enable Controls
Damage or loss through unauthorized entry/access to sole-tenant (i.e., company-only occupied) facilities	Physical access restrictions	Perimeter fence (anti-climb features)
		Single point of entry
		24/7 security staff
		Mechanical or electronic locks
		Visitor controls
	Visual monitoring	CCTV
		Security staff
	Intrusion detection	CCTV monitoring in conjunction with intrusion sensors and security staff
	Anti-piggybacking	Controlled access vestibule (i.e., "mantrap")
		24/7 security staff
		Optical sensor technology in conjunction with CCTV and security staff
	Authentication/ authorization with audit	Dual or positive authentication
		24/7 security staff
		ID display
	Due diligence	Employee/contingent background checks
		Vendor background checks
Damage or loss through unauthorized entry/access to data fields (sole- or multi-tenant facilities)	Physical access restrictions	Mechanical or electronic locks
		Single point of entry
		24/7 security staff
		Signage
	Visual monitoring	CCTV
		Security staff
	Anti-piggybacking (if not at facility perimeter)	Turnstile access
		Controlled access vestibule (i.e., "mantrap")
		24/7 security staff

(Continued)

Table 3.5   Security Standard for Facilities that Process Electronic Information (Data Centers)—*Cont'd*

Threat	Controls	Methods to Enable Controls
Damage or loss by authenticated/authorized individuals inside data fields	Authentication/ authorization with audit	Optical sensor technology in conjunction with CCTV and security staff
		Regular review of CCTV footage to see if piggybacking is occurring
		Badging must occur for each individual entering the data field
		Single-factor authentication (in/out) for sole-tenant facilities; dual or positive authentication for multi-tenant facilities
		24/7 security staff
		Computer room entrance door alarm programmed as "critical" and generates an audible and visual alarm during a "door forced" or "door held" incident resulting in an immediate guard response
		ID display
	Due diligence	Employee background checks
		Vendor background checks
	Controlled equipment deliveries	Pre-authorization
		Visual monitoring/authorized supervision
	Visual monitoring	Extensive internal CCTV coverage
		Security staff monitoring live footage
	Physical access restriction	Unique code for granting access to the data field (distinct from data center access code)

3.4 SECURITY RISK AUDITS

We should be appropriately celebratory in having created a flexible but rigorous risk assessment framework based on security standards. But the world is not static. Threats and associated risk profiles change, so there is an ongoing need to assess and re-assess the security landscape. This might seem like a particularly onerous and laborious task especially for companies with multiple facilities.

The good news is that security standards based on the risk mitigation process that is central to this chapter point to a ready-made auditing methodology. Ultimately, we want to know that facilities and/or security processes are in compliance with a set of risk-based security standards since these standards are the yardstick against which residual risk should be measured.

Not surprisingly, the audit process I am about to reveal is linked to these security standards rooted in the risk mitigation process. Any meaningful security audit process *must* be closely aligned with agreed security standards if the goal is to effectively manage risk, since the standards define the limits on residual risk. To that end, this process must focus on identifying gaps between what is required of the security standard and existing conditions.

Although the goal here is to provide a rigorous process with which to assess and mitigate risk, it is worth reiterating that there is no substitute for judgment here or in evaluating security risk in general. The security standards require that the type and number of methods needed to implement the prescribed controls be specified. This is a judgment call. In the same vein, a proper assessment should specify what controls and methods exist for a particular area or function being evaluated, compare these to what is called for in the standard and determine if there is a gap.

However, it is quite possible that additional controls and methods are required despite the fact that existing mitigation actually comports with the standard as written. Risk profiles change over time. So revalidating the standard using judgment based on a combination of experience and analysis is as necessary as performing the risk assessment. The time frame for the revalidation of standards is also a judgment call. Local risk conditions should be factored into this decision.

Finally, notice that the risk mitigation methodology used to develop standards and audits naturally leads to the generation of security risk metrics. For example, all methods as prescribed by the security standard meet specified performance criteria and associated technical specifications represent a key metric. Other metrics might apply (note that in general metrics measure productivity, effectiveness/efficiency, and risk), but the logic of the risk model driving a mitigation strategy and process linked to a set of standards suggests that compliance with those standards is a key risk indicator.

One example of a possible risk metric derived from security standards is the mean-time-between-failure of retractable bollards. If this figure exceeds the stated technical performance specifications then the measured risk exceeds acceptable levels and mitigation is required.

One possible physical security audit template for facilities is shown in Table 3.6. It is designed to walk the individual conducting the assessment through the process in such a way as to directly assess compliance with the relevant security standard and identify gaps. The audit template is intended to highlight the presence (or absence) of controls and methods as required and is specified by area within a facility to be in direct alignment with the Physical Security Standard for Office Facilities.

Table 3.6 Security Audit/Questionnaire Based on Security Standards

Physical-Security-Audit

Facility Areas	A	B	C	D	E	F	G	H
		List the Controls Currently In Place For Each Audited Area per the Relevant Standard	List the Methods Currently In Place that Enable the Controls Noted in "B" per the Relevant Standard	Do the Methods Noted in "C" Meet the Performance Criteria as Specified in the Relevant Standard and/or Do They Effectively Mitigate Risk? (use the comments section to do document deviations from standards based performance criteria) (Y/N)	If the Answer in "D" is "No", List any Additional Standards-based Controls Recommended –to Mitigate Identified Risk(s)?	List the Standards-based methods (or non standards-based) Recommended to Enable the Additional Controls	Priority of Recommended Enhancements Based on Risk (H/M/L)	Comments
Perimeter and External Environs								
Lobby/Reception Area								
General Office Space								
Technology Rooms								
HVAC Rooms								
Mail/Package Processing Area								
Loading Bay								
Parking								
Wellness/Health Unit Areas								
Child Care Area								
Control Room								
Conference Room								
Stairs and Emergency Egress Point								

3.5 **SECURITY RISK PROGRAM FRAMEWORKS**

As an added bonus, the security risk mitigation process detailed in this chapter can be invoked to establish security program frameworks. These frameworks are useful in ensuring consistency and rigor in applying resources and in effectively managing risk across diverse program areas. By now you will no doubt detect a rather obvious pattern: all programmatic roads appear to lead to the security mitigation process. Keep in mind that the term security program is really shorthand for security risk mitigation program. Therefore, it would be strange indeed if security standards, audits, and program structures are decoupled from the risk mitigation process.

Once again, the set of unique threats is the starting point in establishing a risk-based security program framework, and grouping the most general categories of unique threats represents the first step. This is exceptionally important since all aspects of the program are indexed to such threats. In case you were so tempted, it would be counterproductive to try and list every possible variant of a threat. The challenge is to derive an "irreducible" set of distinct categories under which all threat scenarios can be subsumed.

The next step in accordance with the general approach is to identify the risk factors that enhance the components of risk associated with these general threat categories. Recall that these risk factors drive the required mitigation.

Finally, the controls/mitigation methods that exist to address identified risk factors are the crux of any security risk management strategy/program. As such, meaningful metrics are necessary to gauge productivity, effectiveness and efficiency, and the residual risk. This may be the most challenging aspect of any security program framework. It also adds even more significance to the development of security standards that should be tightly coupled to the risk mitigation process. In view of the inherently defensive nature of security programs, the only meaningful metric available to measure residual risk may be whether a program or facility is in compliance with agreed security standards.

Table 3.7 provides risk-based frameworks for some common security programs, linking unique threats-to-risk factors-to-mitigation methods-to-metrics.

3.6 **SUMMARY**

Understanding security risk begins with accurately identifying and categorizing the unique threats. Evaluating the individual components of risk and associated risk factors is a critical part of a rigorous risk assessment process. Addressing each risk factor in the form of controls,

Table 3.7 Sample Frameworks for Some Common Security Risk Programs

	1	2	3	4
Threats				
	Smoke/fire	Toxic release and/or contamination		
Risk Factors				
Fire safety	1. Code violations or lax codes 2. "Unacceptable" rates of alarms (legitimate and false) 3. Inadequate local fire protection	1. Proximity to facilities that process or store toxic/hazardous materials 2. Open source information regarding leaks, spills, and emissions		
Mitigation Methods				
	1. Code reviews/inspections 2. Alarm tracking/statistics and remediation 3. Drills, training, and awareness 4. Crisis management capabilities 5. Risk assessments	1. Crisis management capabilities 2. Business continuity plans		
Metrics				
	1. Number of false alarms 2. Trend in building evacuation time during drills			
	1	2	3	4
Threats				
	Violence against or disruption of employees/executives in non-company-controlled space	Kidnap and ransom of company employees	Disruption of company-sponsored or -attended events	

	1	2	3	4
People protection				
Risk Factors	1. Negative publicity directed at the firm and/or its executives 2. Company at the focus of political activists	Historical evidence of kidnap/ransom of corporate executives or hostility to international corporations	1. Negative publicity directed at the firm and/or its executives 2. Company at the focus of political activists 3. Public awareness of company-sponsored or -attended events	
Mitigation Methods	1. Executive protection 2. Special vehicles 3. "Advance" security reviews 4. Open source and confidential info sources 5. Risk assessments	1. Executive protection 2. Kidnap and ransom defense 3. Open source and confidential info sources 4. Risk assessments	1. Event security details 2. Open source and confidential info sources 3. Risk assessments	
Metrics	Compliance with standards	Compliance with standards	Compliance with standards	Compliance with standards
Physical Security				
Threats	Violence/criminal activity affecting company employees while in company space or facilities	Thefts of company property and documents	Regulatory or statutory offenses	
Risk Factors	1. Historical evidence of violent crime/terrorism 2. Government alerts and travel advisories re: terrorism or local hostile activity 3. Protests targeting company facilities 4. Poor general economic conditions	1. Corporate downsizing 2. Internal moves 3. Presence of non-vetted personnel on-site	Business/regulatory-restricted entities in physical proximity (i.e., located within the same facility)	

(Continued)

Table 3.7   Sample Frameworks for Some Common Security Risk Programs—*Cont'd*

	Mitigation Methods	
1. Guarding 2. Visual monitoring 3. Police details 4. Radio communications 5. Risk assessments 6. Enhanced façade/window design 7. Explosive detection 8. Open source and confidential info sources 9. Barriers/bollards 10. Risk assessments	1. Authorized entry to company facilities 2. Authenticated entry to corporate facilities 3. Visual monitoring 4. Risk assessments	1. Authorized entry to corporate facilities and restricted space 2. Authenticated entry to corporate facilities and restricted space 3. Visual monitoring 4. Guarding 5. Risk assessments
	Metrics	
1. Compliance with standards 2. Retractable bollard mean time between failures 3. Compliance with standards	1. Trend in number of lost/stolen employee IDs 2. Trend in number of thefts	Trend in the ratio of department/division population to the physical access privileged population

methods, performance criteria, and associated technical specifications represents a flexible and simple but rigorous and inclusive security risk mitigation process.

Effective risk assessments are predicated on developing security standards based on a formal risk mitigation process such as the one presented in this chapter. Some examples of standards were included in this chapter to illustrate the rigor and flexibility afforded by the proposed structure. Performance criteria and associated technical specifications for some key mitigation methods are available in Appendices D to G.

Security standards used to assess risk naturally lead to the establishment of meaningful security metrics. Such metrics can measure residual risk, which is the gap between the existing mitigation and what is called for by the standard. Security audits aligned with the security standards are an integral part of an effective security risk management program and facilitate continuous adjustments to risk mitigation strategies in response to dynamic risk profiles.

Finally, the creation of risk-based security program frameworks are natural by-products of the risk mitigation-security standards-risk audit ensemble as specified herein.

Part **2**

Measuring and Mitigating
Security Risk

Chapter 4

Measuring the likelihood component of security risk

When the personality of a human is involved, exact predictions are hazardous. — McCoy

"The Lights of Zetar," *Star Trek,* **Stardate 5725.6**

Random chance seems to have operated in our favor. Spock
In plain, non-Vulcan English, we've been lucky. McCoy
I believe I said that, Doctor. Spock

"The Doomsday Machine," *Star Trek,* **Stardate 4202.9**

It would be illogical to assume that all conditions remain stable. — Spock

"The Enterprise Incident," *Star Trek,* **Stardate 5027**

4.1 **INTRODUCTION**

An accurate estimate of the likelihood that a threat will be directed against a particular facility or entity would be extremely useful information pursuant to the development of a risk mitigation strategy. Security professionals are often called upon to opine on how likely a threat is. Quite often the answer to such a question is based on intuition gleaned from experience. Is that adequate, and if not, are there methods available to augment intuition? When are these methods applicable?

It is important to be rigorous about assessing the likelihood component of risk, specifically to be precise about exactly what is predicted and how this prediction is made. In particular, accurate distinctions must be made between the likelihood of incidents for which the rate of occurrence can be characterized by statistical processes and those that cannot. Relying on the past to predict the future can be tricky for some threats since conditions change with time. The probability of a future occurrence may have little connection to the past.

For security risk incidents that occur at random, established statistical methods can help facilitate estimates on the likelihood component of risk assuming certain conditions are satisfied. Other types of security incidents, especially those where sufficient data are lacking, are less amenable to quantitative estimates on likelihood. This chapter discusses both types and provides guidance on when specific techniques are applicable in assessing the likelihood component of risk.

4.2 **LIKELIHOOD OR POTENTIAL FOR RISK?**

Reliable estimates of the likelihood of an event for non-random processes should be based on a history of similar events that have occurred under similar conditions. This is how insurance companies make money. For example, the likelihood that a teenager will be involved in an automobile accident while driving is based on the rate of accidents experienced by his peers. Actuaries then develop a model based on the data from accidents committed by legions of wayward offspring. From this they calculate expected financial loss. As you can probably guess, automobile insurance premiums for teenage drivers are expensive and it does not take a brain surgeon (or even a security consultant) to understand why.

To be fair each teenager is different and remarkably some might even be responsible drivers. But statistics point to an established pattern of bad behavior as a group with a record of accident-related financial loss. Estimates on the likelihood of an accident involving your teenage son or daughter are based on generalizations from historical data that presumably have proven relatively stable over time. The unfortunate fact of life is that responsible teenage drivers are statistical hostages to their less responsible peers.

People understandably seek reliable information on the likelihood of the next security or safety-related incident such as a terrorist attack, criminal act, earthquake, etc. Although crystal balls are not typically part of the security professional's arsenal, there are useful techniques that can be applied assuming appropriate conditions are satisfied.

It is important to distinguish between likelihood and potential when one speaks of the probability of a future security incident. The term likelihood is often applied quite broadly in this context and can be misleading due to an implied level of precision. For many types of security incidents the spectrum of possible outcomes is nearly infinite, unknown, and/or varies with time. In these cases great care must be exercised in quoting an exact probability and then only with appropriate caveats. In Section 4.5 I provide an example of an actuarial model that calculates the probability of loss from building damage due to terrorism. This model is indeed based on a technically rigorous methodology. However, recognize that fancy mathematics notwithstanding, such estimates are inevitably based on subjective opinions when assessing threats characterized by inherently uncertain risk factors.

Estimating the likelihood of a terrorist event is not the same as figuring out the odds of hitting a particular number in dice or the odds of a specific sequence of coin flips. In both these processes we know all the possible outcomes in advance, so the probability of a given outcome represents a fraction of the totality of known outcomes.

Specifically, there are 6 faces to a die and 2 die per pair of dice, so there are 36 possible outcomes ranging in value from 2 to 12 in face value. The probability of hitting a combination corresponding to the number "two" (i.e., "snake eyes") is 1/36 or about 2.7%. Significantly, there is no history to be learned from these dice. Whatever outcomes occurred in the past is irrelevant to future outcomes, but all possible outcomes are known in advance. The outcome of each roll is therefore *independent* of every other roll.

The very notion of "hot" dice is fundamentally flawed and is caused by a lack of appreciation for the occurrence of statistically unlikely outcomes as well as confusion regarding the concept of independence in probability. Random processes can give the illusion of being influenced by unrelated factors and appear to be causal or non-random.

Truly random processes do provide statistically unlikely surprises similar to those regularly experienced at the craps table. Since many of the truly important lessons in life can be gleaned from baseball, I refer the reader once again to the March 30, 2008, *New York Times* Op Ed piece by S. Arbesman and S. H. Strogatz referenced in Chapter 1. The authors performed a Monte Carlo simulation using the actual records of every hitter in every game since the inception of baseball. The Monte Carlo method, developed by scientists working on the Manhattan Project during World War II, uses random number generators and multiple trials applied to models to create probability distributions.

This exercise revealed the rather surprising result that Joe DiMaggio's 56 game hitting streak was not so rare. The longest streak in this virtual exercise was an astounding 109 games. Even more surprising was that household names like Hugh Duffy and Willie Keeler were the most likely individuals to achieve this record and not Joltin' Joe.

If all the outcomes of a process are known, and the conditions influencing those outcomes are stable over relevant timescales, one can make a quantitative estimate of the likelihood of occurrence for a future security incident. However, this can be a big "if" when applied to problems in security risk.

For some problems in security risk it is absolutely critical to exploit the lessons of history to estimate the likelihood of future incident occurrences. As a security professional it would be silly and possibly even negligent to mistakenly think that incidents resulting from a clearly biased process were occurring at random. This would be tantamount to ignoring the risk factors that influence the likelihood and vulnerability components of risk. Understanding the risk factors is at the heart of the risk assessment process and should be a principal focus of a security professional.

The problem with real life is that risk factors often change and/or may not be very well understood. This introduces an inherent bias in the process, which is manifested as uncertainty. Typically the more complex the process is, the more uncertain is the likelihood of a specific outcome. The principal exercise for security professionals is to distinguish random from biased security-related processes and to apply judgment modulated by analytic skill in appropriate proportions.

Humans sometimes mistakenly believe that because events in real life happen sequentially or over a short time interval, then such events are causally related. These same people sometimes make assumptions regarding likelihood based on their recollection of history. For example, people consistently think this way when discussing various medical ailments: "my symptoms disappeared when I applied a magnet to my injury" or "my fever broke when I took some of grandma's chicken soup," etc. Although I intend no ill will toward grandmas, and I am a firm believer in the palliative effects of chicken soup, relationships between cause and effect can be quite coincidental and therefore illusory.

An apocryphal story that illustrates this point is the one about the man who attempts to carry a bomb aboard an airplane. His plan is foiled by attentive security personnel before takeoff. During questioning the man is asked about his motives. He responds candidly by saying, "Well, I figured the probability of one person having a bomb on an airplane is small, and so the probability that TWO people on the same plane have a bomb is incredibly small!" Hopefully you can now better appreciate the flaw in this logic and avoid implementing a similar security strategy.

4.3 **ESTIMATING THE LIKELIHOOD OF RANDOMLY OCCURRING SECURITY INCIDENTS**

If certain events are independent, then the probability of two such events occurring is the product of their individual probabilities. We can take this as a working definition of statistical independence. The probability of rolling a single die twice in a row and hitting a one each time is $1/6 \times 1/6$ or $1/36$. It is even possible to roll a one 100 times in a row, but the odds of doing so are $(1/6)^{100}$, which is an extremely small number.

The notion of independence is relevant to the likelihood component of risk in security. Suppose as the security director you are asked to develop a five-year business continuity program for a facility and are concerned about maintaining building power in the event of a crisis. As a first step you check on the resilience of the power supplied to your facility. You discover that three electrical generators are used to power your building and that two are required to be simultaneously functional to deliver adequate power.

The generator manufacturer's data show that the probability of a single generator failing in 5 years is 0.1 or 10%. This figure seems high and you intend to speak to the facility manager as soon as the power is restored to your building. The manufacturer presumably tested a number of units to arrive at the 10% failure rate. Note the manufacturer may not have tested the specific unit you purchased, but might have instead tested some statistically significant number of similar units to arrive at a general understanding of risk with respect to this particular make and model. A statistical distribution known as the Weibull distribution is sometimes used to analyze reliability problems in manufacturing.

Testing of this sort is crucial to understanding the likelihood of encountering problems with machinery in general, and you make a mental note to pursue this concept in more detail. However, in the interest of expedience you continue with the calculation to try and understand the likelihood of a generator-induced power failure at your facility.

If there is a total of three generators, each can exist in one of two states: failure or functionality. There are eight possible states for the three generators, and each generator state is independent of one another. If we call the functional state 1 and the failure state 0, we know from the manufacturer's data that the probability of generator failure is 0.1, therefore the probability of functionality is $1 - 0.1 = 0.9$. So a state where the first two generators function and the third one fails (i.e., 110) has a probability 0.081 or 8.1%. Notice that there are three possible states with similar outcomes and exactly the same probability of occurrence: 011,

101, and 110. The list of all possible states for the three generators with associated probabilities is as follows:

Generator States	Probability
111	(0.729)
110	(0.081)
100	(0.009)
101	(0.081)
011	(0.081)
010	(0.009)
001	(0.009)
000	(0.001)

Notice that the sum of the probabilities equals one. This makes perfect sense and could not be otherwise. The sum of the probabilities for *any* distribution equals one by definition. The list shown here represents an example of a binomial probability distribution. The "bi" in binomial refers to the two possible outcomes of the defining states, which are specified in this case as failure and functionality.

Another famous example of a binomial distribution is the one that results from performing coin flips. In this case the probability of the two states, heads and tails, is the same (i.e., 0.5) for either outcome if using an unbiased coin. A general expression for the binomial probability density function is given by

$$P(x) = \frac{n! \, p^x (1-p)^{n-x}}{x!(n-x)!} \qquad \text{for } x = 0, 1, 2 \ldots n \qquad (4.1)$$

Here n represents the number of trials for independent events that each has the probability p of occurring. $P(x)$ is the probability of x successes out of n trials. The exclamation point represents the multiplication of integers in a descending sequence (e.g., $4! = 4 \times 3 \times 2 \times 1$) and is known as a factorial.

What does the distribution look like for the outcomes associated with flipping two coins? The four possible outcomes are HH, HT, TH, and TT. As noted above, if the coin is unbiased then the probability of a single flip of heads = probability of a single flip of tails = 0.5. So two heads will appear with a probability of $0.5 \times 0.5 = 0.25$, the probability of one head is $0.25 + 0.25$ (i.e., HT and TH), and 0.25 is the probability of zero heads (i.e., TT). Applying the formula for $P(x)$ we get the same results:

$$\text{Probability of zero heads} = P(0) = \frac{2!(0.5)^0(1-0.5)^{2-0}}{0!(2-0)!}$$

$$= 0.25 (\text{note: } 0! = 1) \qquad (4.2)$$

$$\text{Probability of one head} = P(1) = \frac{2!(0.5)^2(1-0.5)^{2-1}}{1!(2-1)!} = 0.50 \qquad (4.3)$$

$$\text{Probability of two heads} = P(2) = \frac{2!(0.5)^2(1-0.5)^{2-2}}{2!(2-2)!} = 0.25 \qquad (4.4)$$

An example of a binomial distribution is shown in Figure 4.1.

Later we will be introducing two other probability distributions relevant to security analyses, the Poisson and the normal or Gaussian distributions.

But what does all this have to do with security? Returning to the generator problem, the goal is to determine the likelihood of a building power outage due to failed generators. This could be an important issue for security professionals who must assess the risk associated with threats affecting business continuity such as natural disasters.

We happen to know that two functional generators are necessary to ensure there is no business disruption; i.e., *any* two generators can be working and our facility continues to maintain power. Based on the spectrum of possible generator states previously listed, there are three states yielding two functional generators and each has an 8.1% chance of occurring.

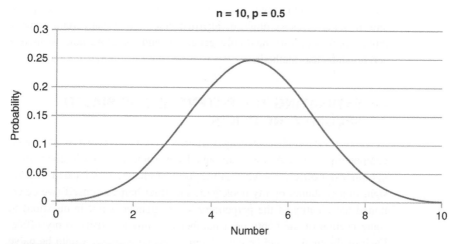

■ **FIGURE 4.1** Binomial distribution, n-100, p=0.5.

There is one state where all three generators are functional (i.e., 111) and this has a 72.9% chance of occurring since each generator operating independently has a 90% chance of working. Therefore, the likelihood of the building maintaining adequate generator power over a 5-year period is $(0.081 \times 3) + (0.729 \times 1) = 0.972$ or 97.2%. Conversely, there is a 2.8% probability of losing power due to two or more simultaneous generator failures during any 5-year period. Notice we are not interested in the likelihood that a *specific* pair of generators will fail (i.e., $0.1 \times 0.1 = 0.01 = 1\%$).

The manufacturer's data revealed a 0.1 probability of a generator failing under normal operating conditions in a 5-year period. Presumably this number was achieved by extensive testing on the part of the manufacturer. Unfortunately, testing necessary to develop an analogous figure of merit for security threats is often impractical. On the other hand, security equipment such as cameras, turnstiles, retractable bollards etc. should include manufacturing data that includes the mean time between failures. Recall metrics of this type were cited in Chapter 3 as one means to assess risk. But how does one accurately and reliably determine the probability of a security *incident* in the face of a dynamic and only qualitatively understood threat?

The generator manufacturer exploited the fact that test conditions were relatively stable over time to arrive at the 10% probability of failure. Indeed if external conditions that influence the risk factors for a threat remain stable over appropriate timescales then we can use historical data to generate a distribution of outcomes. This would allow us to make estimates of the likelihood of a future incident in direct analogy with the generator problem. In some cases we might be justified in assuming that certain security threat "processes" are randomly generated and therefore able to invoke certain statistical methods.

4.4 ESTIMATING THE POTENTIAL FOR BIASED SECURITY INCIDENTS

Let's examine another security scenario with respect to the likelihood component of risk. A few years ago I had some petty cash stolen from my *unlocked* office. The cash in the form of change was located in plain sight in a container on my bookshelf. The thief helped himself (we determined the identity of the perpetrator so the gender is not in question) to some portion of the booty on a number of uninvited visits to my office. Each of the thefts could be considered independent but it would be naive to think the thefts were not correlated.

I could have presumably influenced the scenario by locking my office and/ or removing the cash. These simple acts would have significantly reduced the vulnerability component of risk. In fact, such preemptive measures might have affected the potential for future thefts as well. But in this case the risk conditions remained constant over time as I stubbornly kept my office unlocked and invitingly left the money in plain sight. Can one compute the likelihood that there will be future thefts? Is this calculation similar to a game of dice?

As discussed previously, we know the exact set of possible outcomes when we roll a pair of dice: the probability of rolling a specific number is just the number of dice combinations that can produce that number divided by 36. In exact analogy, we require an understanding of the set of outcomes for my office theft scenario. Consider the probability of throwing a number in dice if one possible outcome is a "no throw" *and* this outcome depends on the mood or condition of the thrower. It is also possible that our repentant thief developed a conscience and renounced thievery forever. Hope springs eternal.

Alternatively the thief could have moved on to a softer target, although I'm not sure how this would be possible in this instance. My guess is the "perp" (law enforcement parlance) could not believe his luck in finding a so-called security professional that is so cavalier in leaving money lying around an unlocked office.

If one knew with certainty that an office on a particular floor with one hundred offices was being targeted, then it could be said with absolute assurance that there is a 1 in 100 probability that a specific office will be victimized. In an admittedly ludicrous scenario, suppose there are 10 buildings on campus, and we assume all buildings are equally vulnerable to terrorist attack. If an attack is certain (think of how someone could actually know this tidbit of information), the likelihood that a specific building will be impacted is 1 in 10 or 10%.

Should a security director therefore recommend an increase in the number of buildings and decrease the odds of attack on a specific facility? Diversity is indeed a legitimate means of reducing risk by minimizing its "concentration." It effectively guards against putting all of one's eggs in one basket. Although this strategy is sound from a strictly mathematical perspective, I doubt it will win big points with senior management in this case.

The thefts from my office exemplify the distinction between likelihood and potential for incident occurrence in analyzing security risk. The former applies when a precise calculation can be made based on perfect knowledge of the spectrum of possible incident outcomes. The latter is an estimate driven by varying risk factors that influence the set of potential outcomes and introduces uncertainty in the distribution of those outcomes.

Returning to the thefts from my office, calculating the likelihood of a future theft is impossible but most security professionals would bet money on a return visit by our perp. Although an exact calculation of likelihood is impossible, this does not relieve one from an obligation to apply common sense. In fact our man did return one too many times as he most likely could not resist the temptation of such easy pickings. The combination of greed and dishonesty represent well-known risk factors that unfortunately influence the likelihood component of risk associated with the threat of theft.

So what about a mitigation strategy to address the identified risk factors? If I had simply locked my office or placed the money out of view our petty thief would have probably been discouraged and looked elsewhere. Since this would have been a no-cost solution, the value for money to reduce the vulnerability component of risk seems difficult to question. Suppose the only effective mitigation was to install an expensive alarm system to protect my petty cash? This provides a useful segue to a discussion about loss that is germane to the overall risk assessment process.

One convenient feature of theft from an analytic point of view is that there is usually an established value associated with each stolen item. This allows us to quantify loss and facilitate risk versus gain decisions. In a simple but extreme case in point, one would not necessarily want to protect $100 worth of items with a $1,000,000 alarm system.

But sometimes this risk calculus may not yield such a valid metric. What if an epidemic of petty thefts materialized in a company's global headquarters? This could result in a disruption of the work environment and cause employees to lose confidence in the corporate security apparatus and by extension the company itself. Add to these troubles the perception that the thefts, albeit small in value, have contributed to a drop in the company's global reputation.

In these circumstances, one might be inclined to invest in that $1,000,000 alarm purchase despite the small dollar value associated with each incident. It is incumbent upon the security professional to understand the threat and its potential impact on the organization. The good news is that sometimes a few simple statistical parameters will yield insights in this regard.

Finally, one must be able to articulate the precise scenario when making assessments of risk. Site-specific conditions might include risk factors that significantly bias the process. For example, the fact that the historical death rate from lightning strikes in Manhattan is extremely low should be small comfort to someone clinging to the antenna at the top of the Empire State Building during a violent thunderstorm.

4.5 **AVERAGES AND DEVIATIONS**

What parameters are important to evaluate the likelihood component of risk? The concept of an average, also known as the mean of a distribution, is critical to interpreting data and to assessing security risk. We may be interested in the average value of stolen items, the average number of thefts per period of time, the average value of recovered items, the average number of thefts per building, the average number of terrorist incidents, etc. An average is a way of expressing a general characteristic of a group, although no single member of the group may actually be characterized in this way.

The average value of a set of entities is equal to the sum of the values of each entity in the group divided by the total number of entities in the group. If a building has experienced thefts of items worth $100, $120, and $150, the total value of stolen goods equals $370. The average value of thefts is determined by dividing this sum by the total number of thefts, or $370/3 = $123.33. Notice that none of the three individual thefts matches the average value; the average of the deviation from the average is about $21.00 or 6% of the total.

Now consider four incidents of theft valued at $10, $10, $1000, and $1000, respectively. As before, the average is computed by adding the individual values of theft and dividing by the number of thefts, which in this case is four. This yields an average theft of $505, but in this case the average deviation from the average is $495.00 or 25% of the total.

Note again that no single incident of theft comes close to the computed average value and that the "spread" or average deviation from the average is much broader. In the latter case, merely quoting the average value of stolen items would be providing a limited picture of reality since there was an equal number of very low-value items stolen as high-value ones.

An example that hits close to home is the quotation of average apartment prices in the rarefied world of Manhattan real estate. It turns out that there were two buildings in Manhattan that have profoundly influenced the average value of real estate sales. In fact, they skewed quoted results for the total population of sold apartments. For that reason even real estate brokers, a group not generally noted for transparency in reporting, sometimes quoted two averages with respect to recent sales; namely, one average that included sales figures for these two buildings and one that did not.

These simple examples highlight the deficiency of quoting only averages when describing a distribution of security incidents or any results for that matter. To overcome this problem, the standard deviation and variance of a distribution are useful. The variance, sometimes referred to as the second moment about the mean, reveals the spread about the computed average of the distribution and is expressed as the square of the standard deviation as follows:

$$\sigma^2 = 1/n \sum_{i=1}^{n} (x_i - X_{ave})^2 \qquad (4.5)$$

The upper case Greek letter sigma, Σ, indicates a summation. The previous formula tells us that the variance is calculated by first subtracting the average from each point in the distribution, squaring the result of each subtraction, and summing the total number of subtractions and dividing this total by n (the total number of points in the distribution). The standard deviation is by definition equal to the square root of the variance and is represented by the lower case Greek letter sigma, σ. If thefts are assumed to be normally distributed where the average of the distribution is equal to $100 and the standard deviation is $10, then by definition the majority of thefts (i.e., 68%) will fall between $100 − $10 = $90 and $100 + $10 = $110. This can be useful information since the spread in the values of thefts tells a lot about each theft "process."

Computing the standard deviation of a distribution is easy when there are only a small number of values to consider. The details are less important than knowing what a standard deviation represents. Many inexpensive calculators can compute averages and standard deviations. The Microsoft application Excel performs averages and standard deviations quite easily as well. The point is to realize that specifying only the average of a distribution is often insufficient and can even be misleading.

This is not to say that knowing the average of a distribution is not useful information. Examining how the average number changes with time allows for statements about the likelihood of future incidents for similar periods. If the average number of thefts in the month of December jumped by 50% in the past 5 years it might be wise to investigate what else is happening that month in or near your facility. In arguing for additional security resources one could state with certainty that December historically represents a banner month for thefts. Therefore, based on the assumption that history tends to repeat itself under similar conditions, the likelihood of a similar statistical experience in the coming December is presumed to be high.

Recall the previous real-life example of the thief who stole money from my office. In that case we had good reason to believe there was a correlation among incidents and an inherent bias in the process of "generating" (perpetrating is probably the better term here) thefts. But what if we assume that the number of thefts on campus was a normally distributed random variable? Can we use the properties of a normal distribution to highlight enhanced risk?

For normal distributions which will be introduced more formally below, the standard deviation is proportional to the square root of the total number of the total population measured. What if you observe that the number of thefts greatly exceeds the square root of that total as measured on a monthly basis? This may be a hint that "routine" statistical fluctuations are not at work. In the interest of continued employment it is worth mentioning that even if the number of thefts is within statistical expectations it is not a justification for inaction.

One caveat is that the precision of a calculation of standard deviation is dependent on the total number of incidents. This is both good and bad news. Large numbers of thefts enhance statistical accuracy with the obvious downside to career aspirations. It is fair to say that enhancing the number of thefts (or any security incident) just to achieve greater statistical confidence is generally a bad idea. Although I am sure this point is completely obvious, it is worth a little math just to show how foolhardy such a strategy would be.

Suppose we require that the precision of our statistical "measurement" does not exceed 10% of the total number of incidents, N. This requirement immediately imposes a minimum value for the number of incidents. What is that value? We now know that for a random process (I assume theft is a normally distributed random variable) the standard deviation of the distribution of incidents is equal to the square root of the total number of incidents, N. So, $\sigma = \sqrt{N}$. But we also require that $\sigma = (1/10)N$ or 10% of the total number of thefts.

Therefore, $\sqrt{N} = (1/10)N$. Squaring both sides we see that N must be greater than or equal to 100 to achieve a 10% level of precision. If 1% precision is desired then it turns out N = 10,000. That represents a rather large sample space of thefts. We see that the required precision scales as the square of the number of incidents, which reveals an inherent limitation of small data sets.

Let's say you are interested in characterizing thefts in your headquarters building over the previous year. One way of examining thefts would be to specify the number of thefts for that time period versus the loss corresponding to each theft. We could plot this distribution and then see what it tells us about the average value of loss as well as the extreme values. We might make assumptions about the future distribution of theft incidents and a reasonable assumption would be that the distribution should not change much if conditions remain stable over time.

As the number of thefts increases, a famous law of statistics states that the shape of the distribution will approach a bell curve. This curve, so-named because of its bell shape, is also known as a normal or Gaussian distribution. The latter name derives from its "inventor," the great German mathematician and physicist Karl Friedrich Gauss (1777–1855). The bell curve is known to many of us from school days where it was used to characterize the distribution of examination scores. The phrase "grading on a curve" implies the use of a probability distribution to peg the grading scale to the average and standard deviation of a distribution characterizing the group performance.

A normal distribution represents a form of exponential distribution since it is a function of the exponential, e. It is similar in shape to the Poisson distribution, which we will encounter shortly. The mathematical formula for the standard normal distribution of a random variable x, which by definition has a mean of zero and standard deviation of one, is graphically depicted in Figure 4.2:

$$y = \frac{1}{\sqrt{2\pi}} e^{-x^2/2} \tag{4.3}$$

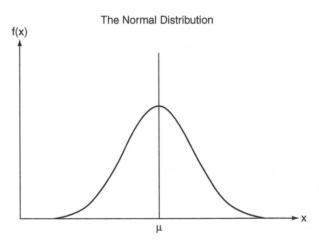

■ **FIGURE 4.2** Normal or Gaussian probability distribution. *From http://www.itl.nist.gov/div898/handbook/ pmc/section5/pmc51.htm.*

A normal distribution yields specific information about the population it is characterizing. The width of the curve is the spread or variance about the mean. In particular, and this holds true for *all* normal distributions, one standard deviation out from the average subsumes 68% of all points under the curve. Two standard deviations include 95% of the points under the curve, and three standard deviations from the mean translate to 99.7% of the entire population. The percentages associated with each of the three standard deviations for normal distributions are useful facts to remember.

Let's make the reasonable assumption that IQ is a random variable and is normally distributed among a large population of individuals. The mean IQ score is 100 and three standard deviations from the mean are scores registering between 55 and 145. This tells us that if a single individual is chosen at random from the population, the probability of that person having an IQ between 55 and 145 is 99.7%. As I noted previously in discussing thefts, both the mean and standard deviation of a distribution are often necessary to appreciate the details of statistical representations of a population.

Individuals routinely make decisions by invoking subconscious if informal distributions of potential outcomes every day. If you are scheduled to visit your mother-in-law in Connecticut for a 6 p.m. dinner and are departing from Manhattan, an estimate of the average time it takes to drive there at that hour is often based on experience. One also probably conjures up the extreme scenarios given risk factors such as time of day, weather, the preponderance of state police, etc. These correspond to the extremes or "tails" of the imagined distribution of travel times.

This information is used to decide on the appropriate departure time to arrive in time for dinner. In this decision process, the implicit assumption is that conditions affecting transit time such as traffic patterns tend to repeat themselves and produce relatively predictable outcomes. I often perform this type of reasoning and it points to the perfectly reasonable strategy of using historical information to estimate likelihood in scenarios where conditions are relatively stable.

To cite another example, between thirty and forty thousand runners compete in the New York City Marathon each year. This represents a great opportunity for robust statistical analyses and Web sites detailing the results provide fodder for those so inclined. It is a good bet that 99% of all runners will register times between 2 and 10 hours. Some important marathon conditions do not change such as the length of the race. Important variables affecting the outcome include the topography of the course, the field of competitors (usually biased by the prestige of the event and the prize money), and the weather.

But one cannot be certain about the outcome. It is remotely possible that someone might finish the race in less than two hours and improve the world record by nearly 5% (i.e., beating the current record of 2 hours 3 minutes and 59 seconds).

This is statistically unlikely but not impossible. You may recall Bob Beamon and his 1968 Olympic performance in the long jump in Mexico City. In a single jump he bettered the world record by an astounding 6.6%. Why is this so astounding? The average improvement in the world record for the long jump prior to this effort was 2.5 inches per jump. Beamon's effort exceeded the previous world record by 21.75 inches. Quantum leaps (no pun intended) in human performance represent statistical outliers and world class athletes are just as constrained by statistical limits as the rest of us mortals. But as evidenced in the case of Bob Beamon, big "jumps" do indeed occur, physically and statistically.

In developing a mitigation strategy for certain threats where sufficient data exist, one might assume event outcomes are normally distributed. Suppose we could assume that the monthly rate of thefts at a facility is a normally distributed random variable. An analysis of the monthly rate of thefts during a 5-year period showed the average number to be 100 with a standard deviation of 10. We know from our previous discussion this implies that 68% of the rate of thefts per month is between 90 and 110, 95% of thefts are between 80 and 120 per month, and 99.7% of reported monthly thefts fall between 70 and 130. The level of precision for the standard deviation of a normally distributed population is a function of the total population as shown previously.

If after performing this analysis your company experienced 140 thefts at this facility during a one-month period, there is probably a need for decisive action. Although one might argue that immediate attention is warranted for a figure far below that. Certain times of the year may tend to skew the average and contribute to a wide dispersion about the mean. If specific times of the year or certain locations consistently showed peak levels of thefts relative to the mean and standard deviation then resources could be deployed accordingly. Knowing the trend in historical incident activity can be very useful in developing a mitigation strategy.

One could perform the same type of analysis for the value of stolen property. If the value per stolen item is believed to be a normally distributed random variable with an average value per item of $1000 and a standard deviation of $200, then the historical likelihood of an item being stolen with a value greater than $800 and less than $1200 is 68%. Note this does *not* say there is a 68% chance a $900 watch will be stolen in the future. There is a 68% likelihood that the value of an item stolen within the last 5 years falls between the aforementioned amounts based on the historical distribution of thefts. If conditions do not change it is reasonable to assume this trend will continue. The distribution could be used to justify specific mitigation measures in terms of the return on investment.

It is sometimes incorrect and even foolish to extrapolate from history when making predictions of the future. However, it would be naive to ignore the presence of trends and patterns to make estimates of likelihood under stable conditions. When interpreted correctly these estimates can be used to develop an informative if approximate risk profile and implement appropriate and proportionate mitigation.

4.6 **ACTUARIAL APPROACHES TO SECURITY RISK**

I mentioned previously that calculating the likelihood component of risk for threats like terrorism is difficult due to unknown and/or changing conditions with the potential to influence the spectrum of security risk outcomes. However, at least two companies of which I am aware, AIR Worldwide and Risk Management Solutions, Inc. (RMS), have developed models for assessing the risk of terrorism.

The AIR version is known as the AIR Terrorism Loss Estimation Model,[1] and their likelihood estimation process uses the so-called DELPHI (as in "Oracle at Delphi") method that relies on expert input to ascertain the likelihood component of risk. According to their Web site, the RMS version is based on a game-theoretic approach for the likelihood estimation process. Both companies have leveraged experience in modeling physical disasters to quantitatively analyze terrorism risk.

I do not know which method yields better results, and in truth I am not sure if this can be determined with statistical confidence. AIR uses historical data on the frequency and magnitude of terrorist incidents in US cities and incorporates input from terrorism experts. They combine these with information regarding defensive security measures to assess the likelihood of property loss. Irrespective of the validity of the model, actuarial approaches to calculating risk provide a means of examining the quantitative effect of specific defensive measures and in theory could facilitate business decisions based on a cost-benefit analysis.

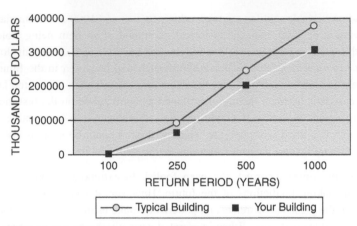

■ **FIGURE 4.3** Probability distribution of property loss due to terrorism.

A graph illustrating the results of a sample actuarial analysis is shown in Figure 4.3 where a specific building in downtown Manhattan ("your building") was compared to similar facilities in the same zip code. The likelihood of property loss due to terrorism using conventional weapons is plotted against the expected magnitude of financial loss. The inverse of the return period is used to indicate likelihood. For example, the loss associated with a return period of 1000 years is equivalent to saying there is a 1/1000 chance per year that expected losses due to terrorism will be equal to or greater than the amount shown on the graph (i.e., about $300M on the vertical axis).

Another way of interpreting the return period is in terms of the time interval over which losses are expected to occur. According to the graph, in 1000 years one can expect property losses from the damage due to terrorism for "your building" to be about $300,000,000 or greater. This is not saying that the probability of a terrorist event is increasing or even changing with time. It is saying that the greater the time *interval* the greater the probability of loss. To understand this better, consider the probability of loss from a terrorist event at a location in the next minute versus the next 1000 years. The probability of an event occurring is roughly the same from minute to minute, but the likelihood of loss increases for greater intervals of time.

The projected loss for the building in question decreases in relation to comparable facilities for increasing return periods; the greater the disparity in the two graphs. To an insurance company, the results of this model indicate that losses due to property damage due to a terrorist attack against "your building" are less likely than with similar buildings. As a result one might argue for lower insurance premiums based on this analysis.

Estimates of the likelihood of terrorist threats rely on an understanding of patterns, demographics, and motives associated with current terrorist groups. Since terrorism is committed by people, and the behavior of people does not always follow predictable patterns, the opinion of trusted experts must be used to gain insight into the potential for incident occurrence. This amounts to weighting various criteria based on expert input and generating a distribution of outcomes. When all is said and done, these experts can only provide opinions and not conclusions based on tested hypotheses. So estimates on the likelihood of terrorism are guesses, albeit educated ones.

That said, a model using explicit criteria and a rational methodology has been posited. In addition, the plot of loss versus return period provides measurement of security risk that might assist in decisions on a mitigation strategy. As an illustration, the decision to add bollards costing $1M could be analyzed using this risk model where the expected loss is evaluated with and without this measure in place. Once a model of this type has been generated resulting in a curve similar to Figure 4.3, it should be possible to make some quantitative assessments of the return on investment of proposed security measures.

Specifically, security strategies for a particular facility could in part be based on the difference in estimated loss both with and without the mitigation to determine value for money. In addition, comparisons made for multiple sites on a campus could yield information on disparities in security and the associated cost of achieving parity. In general, actuarial models of this type could provide a standardized method of evaluating the efficacy of risk mitigation.

4.7 **RANDOMNESS, LOSS, AND EXPECTATION VALUE**

There will be many times when a security manager is faced with a risk decision for threats where no empirical model exists. Is there a way to provide ballpark estimates on likelihood without resorting to extensive research and/or complex modeling schemes?

The answer is a qualified "yes" and characteristically depends on the scenario. To invoke a specific statistical method certain assumptions and conditions must be met. For example, we might assume that risk incidents occur at random. One might argue that the variability in type and effectiveness of the forces used to defend against terrorism coupled with inconsistencies in effectiveness, intent, and the scope of terrorists' efforts combine to produce a random process.*

---

*Losses due to terrorism have been reported to follow a scale-free probability distribution. See "Scale Invariance in Global Terrorism", A. Clauset and M. Young, April 30, 2005.

We might also assume that no correlation exists between event arrivals and that any short interval of time is equally likely to produce an incident as any other. As an aside, the assumption of randomness might not be met with enthusiasm by either perpetrators or defenders as this might impugn their abilities to produce desired outcomes. However, the contention is that their antagonistic efforts *combine* to produce a large number of possible outcomes, so the probability of a particular outcome is inversely proportional to that number. I am sure this contention will be hotly contested by some terrorism experts.

Finally, we might stipulate that the expected number of events is constant over a given time frame. For example, if the rate of occurrence of events averages 10 per week, the expected average number of events in a month will be 40. If this is not the case, we cannot apply our model to time frames of this duration.

If the conditions of randomness, independence, and constant rate of occurrence over a specified time are satisfied, we can assume the arrival of security events behave according to the Poisson process. Despite its name, Poisson has nothing to do with fish, but the underlying theory was developed by the French mathematician, Simeon Denis Poisson (1781–1840). Examples of processes that definitely obey Poisson statistics are the radiation counts emitted from a radioactive source, the number of typos on a typewritten page of material such as this book, cars arriving at a toll booth, and mutations occurring in bacterial reproduction.

The Poisson distribution represents another example of an exponential distribution. As I noted above, to invoke the Poisson process for security events we must assume that the expected number of events does not change over time. The expression for the Poisson function contains a parameter usually characterized by the Greek letter lambda (λ), where λ stands for the expected number of events. Some simplifying qualities of Poisson are that the distribution is completely determined by λ since both the mean and variance of the distribution are equal to λ.

Suppose we are interested in the probability of exactly k events occurring in a five-year period and λ is the rate of events that is expected over a five-year period. The Poisson density function, p(k), is written as follows:

$$p(k) = \frac{e^{-\lambda}\lambda k}{k!} \tag{4.7}$$

Graphs of p(k) with different means are shown in Figure 4.4.

As previously noted, p(k) is solely dependent on the parameter λ, the expected value of event arrivals for a prescribed time period. Therefore,

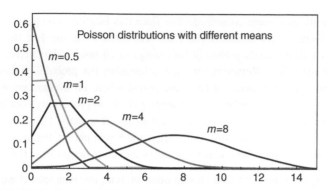

Poisson distributions with different means

$m=0.5$

$m=1$

$m=2$

$m=4$

$m=8$

■ **FIGURE 4.4** The Poisson distribution. *From: http://earthquake.usgs.gov/learn/glossary/?term=Poisson% 20distribution.*

the choice of λ will strongly influence the form of the probability distribution, p(k). Let's apply p(k) to a real-world problem.

A security manager for a corporation in an urban setting has experienced an average of 1.4 catastrophic incidents per year for the previous 5 years. I suspect his job may be in jeopardy but I will continue with the analysis any way. A catastrophic event is defined as an unscheduled building outage that lasts one week or more. The security manager has no reason to believe this rate will change in the next 5 years and he cannot see a pattern or correlation between incidents.

Assuming this rate of occurrence does not change, the expected number of incidents for a 5-year period is $5 \times 1.4 = 7$. What is the probability of the company experiencing *exactly one* incident in the coming five-year period? Note this is not the same as asking the probability of having *at least* one incident during that time, which is equivalent to assessing the probability of one or more incidents. This would require a calculation of the cumulative probability using this same density function.

Although we know λ in the expression for p(k) equals seven in this instance, we cannot determine when any of these seven incidents will occur since they occur randomly. We only know that seven incidents are likely to occur sometime over a five-year period.

The value of k is one in this case since we seek the probability that exactly one incident will occur in a five-year time frame. Recall that the value of the exponential, e, is 2.72. So plugging these values into the expression for p(k) above, where $1! = 1$, we get

$$p(1) = \frac{e^{-7}7^1}{1} = 0.006 = 0.6\% \tag{4.8}$$

This result may seem counterintuitive since this fictitious company historically averages 1.4 incidents per year. Yet for k = 1, the fact that p(k) = 0.6% tells us that the probability of seeing *exactly* one incident in five years is extremely low. Remember we are calculating the probability that one incident will occur during a five-year period where on average we expect there to be seven. Similarly, at the other end of the probability continuum, p(k) for k = 17 tells us that the probability of exactly 17 catastrophic incidents occurring is 0.006% during the same 5-year period. Figure 4.5 plots p(k) for λ = 7 and where k, the number of incidents ranges from 0 to 17.

p(k) is quite sensitive to the chosen time interval. For example, if we consider a two-year time frame, and the rate of incidents is again 1.4 per year, then the expected number of events is 2 × 1.4 = 2.8. If we are now interested in knowing the probability that exactly one event will occur in a two-year period, we may use the expression for p(k) again, but this time with λ = 2.8. This yields a probability of 17% that exactly one incident will occur in a 2-year period. Contrast this number with the probability that exactly one incident will occur in a 5-year period with the same rate per year, or 0.6%.

It must be re-emphasized that using a Poisson distribution to determine frequency or likelihood is based on the assumption that the expected number of incident in a given time period does not change. In the above example, if the yearly rate of event occurrence was not stable then we could not use a fixed λ.

In an example of a real-life situation, I applied this technique to analyze stairwell doors that were not closing properly. This issue was cited as a safety hazard because stairwells must remain free of smoke in the event of a fire. If a door remained unlatched there was a potential for the principal building egress route to become impassable in an emergency.

I assumed that door closer failures were occurring because of random mechanical failures. Maintenance data revealed that the rate of failure was relatively constant over time frames of interest. Applying Poisson to

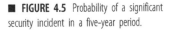

■ **FIGURE 4.5** Probability of a significant security incident in a five-year period.

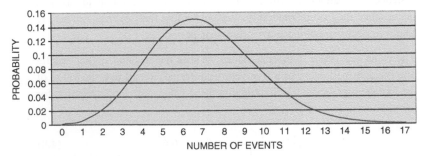

this problem yielded a probability for the number of non-operative doors. This in turn facilitated a more exact appreciation of risk and helped predict the required level of resources to commit to remediate the problem.

The Poisson process offers a means of quantifying the likelihood component of risk for randomly occurring incidents. We can also use this probability distribution to estimate exposure to financial loss when such data are available. The combination of information on the likelihood of security incidents and loss yields a more complete risk profile and hence provides a stronger foundation for security decisions.

Calculating expectation values of security incidents represents a useful element of a risk-based strategy. Another analogy with coin flips is illustrative although coins and dice should probably not be used as security decision tools. Consider a fair coin with two equally likely outcomes of flips, heads and tails. Suppose you were betting on these flips where a toss of heads caused you to win $100 and a toss of tails generates a $50 loss. There are a total of three flips. What is the distribution of expectation values for this process?

The expectation value is defined as the product of the probability of an event with the value of that event. In the case of the three coin flip events, the following represents the distribution of possible outcomes with the expectation value shown in parentheses:

$$
\begin{array}{l}
\text{HHH } (+\$37.50) \\
\text{HTH } (+\$56.25) \\
\text{HTT } (\$0) \\
\text{TTT } (-\$18.75) \\
\text{TTH } (\$0) \\
\text{THH } (+\$56.25) \\
\text{THT } (\$0) \\
\text{HHT } (+\$56.25)
\end{array}
\tag{4.9}
$$

The outcome HHH would yield a $300 payoff (i.e., $100 + $100 + $100) with a 1/8 probability of occurrence. Therefore the expectation value is $300 × 1/8 = $37.50. The outcomes THH, HTH, and HHT all yield the outcome of $150 and cumulatively account for 3/8 of the possible outcomes. Therefore the expectation is calculated as $150 × 3/8 = $56.25. The other expectation values are determined in the exact same way. When losses from security incidents can be assigned probabilities then expectation values can be computed. Such a scheme could be useful in developing a cost-effective mitigation strategy.

Recall that the graph displayed in Figure 4.5 represents a probability distribution for catastrophic incidents assumed to obey Poisson statistics. Suppose you were faced with a decision of whether or not to build a new building and needed to know the financial exposure to catastrophic incidents. Consider that each incident costs your firm an average of $1M for repairs and the cost of a new building is $5M (cheap!).

The probability of ten catastrophic incidents occurring in a 5-year period is about 7% according to the Poisson model. Ten building catastrophes represent a cost of $10M at $1M per building. Therefore the anticipated financial risk or expectation value of loss is equal to 0.07 × $10M or $700K. Figure 4.6 illustrates the results of a highly simplified analysis of financial risk based on the probabilities generated via a calculation of the expectation value. There is roughly a 14% likelihood that losses will total $1M and this is the most likely expectation of loss. A decision to build or not build may now be made by comparing the financial risk associated with continuing to repair old facilities versus the cost of constructing a new facility that may prove more resilient.

It may appear that by using the Poisson process I have skirted the requirement for a valid behavioral model. In fact, the assumptions used to invoke Poisson *are* behavioral statements and are incorporated in the density function, p(k). The Poisson process can be applied to any type of security incident for which the aforementioned conditions apply.

We began our discussion on likelihood by noting a requirement for techniques to assess the likelihood of future security incidents. The ability to make valid statements about the future expressed as a probability represents a powerful resource. Such estimates of the likelihood component of risk would be immensely valuable to any security professional. As I have now noted many times, evaluating the likelihood of a future threat is predicated on the use of a valid model and exploiting a particular model is determined by scenario-specific conditions.

Consider a decision on whether or not to utilize biometric access control technology. Such technologies include fingerprint readers, hand geometry readers, iris recognition, etc. These have emerged as viable means of ensuring "positive" authentication of individual identity. Approximate and

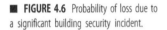

FIGURE 4.6 Probability of loss due to a significant building security incident.

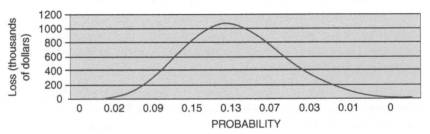

anecdotal error rate (i.e., false positives plus false negatives) for the iris scan technique is about 1/175,000.[*] In other words, a company who deployed this technology would expect either a false rejection rate (the biometric falsely rejected someone and denied physical access) or false acceptance rate (the biometric falsely accepted someone and granted physical access) once every 175,000 times the device was used.

In terms of the risk of unauthorized access or falsely accepting an individual the situation is actually better than the aforementioned numbers alone suggest. This is assuming the biometric is used in conjunction with a card or some form of token. For the system to admit someone in error and be a legitimate risk event a false positive must be evaluated as a conditional probability: the biometric must fail *and* the individual must be presenting a false credential. The joint probability of the credential validation system and the biometric failing should be very small.

Let's examine the likelihood of error for the biometric security device in more detail. The error rate for iris recognition technology seems quite low on first inspection. Consider a building population of 12,000 individuals. Assume we have perfect attendance and each individual feverishly works at their desk all day without leaving the building until it is time to go home. Therefore, every person uses his badge to enter the building once per day. Let's also assume we have one iris recognition system per portal with each portal used equally by incoming staff. There are a total of ten portals.

The number of biometric-induced errors per day per portal $= 12,000$ authentication events/day $\times$ 1/10 $\times$ 1 error/175,000 events $= 12$ errors/ 175 days $= 0.0069$ errors/day-portal. Therefore, 0.0069 errors/day $\times$ 1 day $= \lambda$, the rate of occurrence as denoted in the Poisson density function. According to Poisson, the probability of k error-induced incidents is

$$P(k) = e^{-\lambda}\lambda^k/k! \tag{4.10}$$

With respect to iris recognition technology, we see that the probability of zero errors in one day is about 99%, which makes sense given the error rate λ is so small. However, the high rate of usage yields a non-zero probability of experiencing precisely one error to the tune of 1.4% each day. The probability of two errors in one day is about 0.5% and the likelihood of more errors in a one-day period falls off precipitously from there.

---

[*]Statements on error rates for iris recognition systems vary. Vendors sometimes report a single statistic that encompasses both false rejection rate and false acceptance rate but the two can vary significantly. One vendor, Rehoboth, indicated that their device had a false rejection rate of 1/1000 and a false acceptance rate of about 1/1,000,000.

The cumulative probability of an individual scanner experiencing exactly one or two errors in a day is 1.9%. The probability of experiencing either one or two incidents at a single turnstile in a 10-business-day cycle is 6.6%.

The moral here is that some sort of performance analysis should be conducted in advance of deploying security technology. The security professional must understand both the likelihood of an error and the implications of experiencing such an error in the intended environment. Ultimately a judgment is required to determine if the error rate is acceptable. In this case an error translates into either admitting an unauthorized person or more likely rejecting someone who has proper authorization to enter the facility.

4.8 **FINANCIAL RISK**

In the financial world, analysts devote considerable resources to evaluating the so-called value-at-risk (VaR). Although not exactly applicable to problems in security risk, the VaR offers lessons in understanding the likelihood and vulnerability components of security risk. Figure 4.7 illustrates this concept.

The probability distribution, which in this case takes the form of a so-called log-normal distribution, denotes losses on the horizontal axis versus the relative frequency of occurrence along the vertical axis. The extreme right

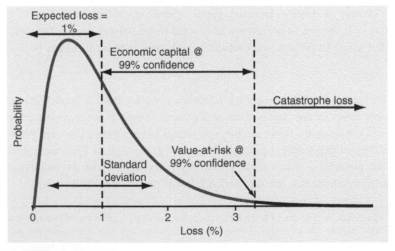

■ **FIGURE 4.7** Value-at-risk.

portion of the curve (i.e., portfolio losses greater than 3%) is considered the VaR in worst case. In other words, there is a 1 in 100 chance that losses will be 3% or greater of the portfolio value. The total portfolio losses can be expected to exceed 3% on 1 out of 100 trading days.

Analysts are able to generate this curve because they have developed a financial model that (they hope) characterizes the value of assets in a port-folio based on factors that influence pricing. Because the value of stocks, bonds, options, etc., is measurable in tangible units (i.e., denominations of money), and financial analysts have faith in their model, they can determine their maximum exposure to portfolio loss with a specified confidence level.

Notice that VaR does not specify the maximum exposure to loss. If you are really unskilled or just plain unlucky you can lose the whole portfolio, and that is not what VaR is measuring. VaR establishes confidence limits for potential loss. Recall our three coin flips where heads yields a $100 gain and tails results in a $50 loss. Although the coin flip example represents a highly simplified model, the same principle is used to develop a financial strategy.

Clearly, the question of whether loss models so constructed are accurate in assessing risk is a legitimate one and serves to reinforce the fact that just because an analysis is quantitative does not mean life will turn out that way. There are indeed other lessons to be derived from the 2008–2009 global financial meltdown, but one of them is that genuinely rare events do occur.

Security professionals desire a similar tool to analyze their version of risk. An important caveat is that security risk assessments should not necessar-ily be focused on the monetary value of assets. The security professional must also protect human lives. Moreover, assets like technology and infra-structure-related items are usually not objects to be traded or sold like financial securities. However, estimates of potential loss based on appro-priate statistical models of security incidents would likely be useful input for decision makers.

4.9 **SUMMARY**

The fact that incidents resulting from security-related processes occur at random can be used to exploit well-known statistical models in estimating the likelihood component of risk. Intuition derived from experience can play an important role in assessing the potential for occurrence of future security incidents that are biased relative to the spectrum of possible out-comes as a result of changing risk factors.

Table 4.1 Summary of Methods to Calculate the Likelihood Component of Risk

	Likelihood Calculation	Restrictions/Caveats	Output
Randomly occurring security incidents and time intervals are independent	Poisson process	Requires constant rate of incident occurrence	Likelihood of an exact number of incidents in a given time interval
Historical record of security incidents	Normal distribution	Valid for a large number of incidents and stable operational conditions	Confidence intervals for a normally distributed random variable
Security incidents resulting in financial loss	Actuarial methods	Validation is difficult	Return period (probability of loss exceeds a specified amount)

In the case of thefts from my office, we observed that the risk was influenced by the risk factors of greed and no locks on the door, i.e., the persistence of conditions that enhanced the likelihood and vulnerability components of risk. These factors biased the distribution of outcomes. As we learned in Chapter 3, mitigation strategies are based on an evaluation of these risk factors.

Specifically, when security incidents are believed to occur at random, a Poisson distribution may be applicable to determine the likelihood of a specific number of incidents in a prescribed time period. The Poisson distribution happens to be the statistical distribution of choice in determining likelihood in the Loss Distribution Approach, which is a popular approach in calculating operational risk.[2] If large numbers of incidents are recorded, the normal distribution is useful in estimating the likelihood of future incidents assuming conditions are expected to remain relatively stable. In terms of characterizing a population of security incidents, both the average and standard deviation or the "spread" of probability distributions are important in developing an accurate picture of risk.

Estimates of risk in terms of exposure to financial loss are sometimes possible through the use of actuarial models where the return on investment relative to specific forms of mitigation under consideration could be evaluated.

Table 4.1 is a summary of some of the methods used to estimate the likelihood component of risk:

REFERENCES

1. www.air-worldwide.com.
2. Aue F, Kalkbrener M. *LDA at Work*. Deutsche Bank AG, November, 2006.

Chapter 5

Measuring the vulnerability component of security risk

Physical laws simply cannot be ignored. Existence cannot be without them. — Spock

"Specter of the Gun," *Star Trek*, **Stardate 4383.5**

5.1 **INTRODUCTION**

A rigorous examination of security risk entails analyzing each component of the fundamental expression of risk introduced in Chapter 1. In this chapter the vulnerability component of risk is investigated in detail. The good news here is that for some threats this component is amenable to a quantitative approach to analyzing risk. This is because some of these threats are influenced by physical processes and/or described by physical quantities that obey well-established natural laws.

Estimating and minimizing the vulnerability component of risk are key elements of a mitigation strategy. Ideally this should occur in conjunction with estimating the likelihood component of risk discussed in Chapter 4. Although the same physical phenomena can sometimes affect the vulnerability component of risk for disparate threats, I usually opt to discuss unique threats and their associated risk components separately. When assessing risk it is important to understand the specific risk factors in each case as these strongly influence the required mitigation.

Another important comment should be made at the outset. Evaluating risk is often an exercise in approximation. Because the intentions and methods of adversaries are usually not known except possibly by the adversaries themselves, one must often rely on reasonable estimates of the parameters that affect the vulnerability component of risk. Through sleight of hand

and/or self-delusion it is possible to make the numbers justify almost any position and convince yourself and others that a problem does or does not exist.

However, if realistic values of key parameters are applied to physical quantities that are known to govern security-related phenomena, a ballpark result is both a reasonable and useful outcome. In the spirit of discovery and the search for truth, when assessing risk it is always a good idea to obtain a "reality check" by seeking confirmation of assumptions and calculations from independent, knowledgeable, and trusted sources. Finally, applying established physical principles and rules of thumb to arrive at estimates of the vulnerability component of risk should be viewed as one piece of the assessment puzzle with the goal of developing an effective and proportionate risk mitigation strategy.

5.2 VULNERABILITY TO INFORMATION LOSS THROUGH UNAUTHORIZED SIGNAL DETECTION

The key responsibility of a security professional is to protect the people, assets, and reputation of the institution for which he works. One of the most important assets of a company is proprietary and/or sensitive information about clients, strategy, and employees. The creation, storage, transmission, and destruction of company information are ongoing processes and the compromise of any of these can have damaging consequences.

Let's set the stage with the following scenario: a radio frequency signal from a local area network (LAN) is radiating unencrypted, company-proprietary information. Suppose that at a distance of 100 meters from the LAN access point (i.e., the transmitting and receiving element) the signal strength exactly equals the strength of the noise signal in the atmosphere across the signal bandwidth.

The vulnerability to the threat of detection is theoretically zero at that distance since the information carrying signal power is equal to the noise power. Engineers often refer to the signal power relative to noise power as the signal-to-noise (S/N) ratio, and it is an extremely important parameter in estimating vulnerability to the threat of information loss through unauthorized signal detection and subsequent demodulation and/or decryption. In the case where the magnitude of signal and noise power is equal, the S/N ratio is unity.

Now consider that the unencrypted signal from your wireless LAN is radiating energy 500 times the magnitude of the noise signal, which is equivalent to saying there is a S/N ratio of 500. Although this figure may seem alarming, we need to determine what this actually means in terms of real detection capabilities by a potential adversary.

Although the S/N ratio is a legitimate risk metric, in practical terms large ratios would not make a big difference to a would-be eavesdropper once a certain threshold ratio is achieved. The vulnerability to the threat of signal detection is the magnitude of signal power that exceeds the dominant source of noise power in that frequency range and across the signal bandwidth. It is therefore very useful to determine the S/N ratio at a point in space to assess vulnerability to detection by an adversary who may be situated at that very spot.

Note that the value of information contained in the signal is unrelated to its vulnerability to detection. The vulnerability here refers to susceptibility to the threat of unauthorized signal detection and misuse of the information contained in that signal. We are not concerned with the financial loss associated with any information so disclosed, although this is clearly an important consideration. If the information in the signal is worthless (i.e., the impact component of risk is zero) then presumably there would be little incentive to protect it.

Unauthorized electromagnetic signal detection and misuse is one example of information loss and is really a form of theft. Contrary to popular belief and despite its dramatization in the popular media, electronic interception would not necessarily be the first explanation for the loss of sensitive information. Ample opportunities for these types of incidents exist through simple acts of carelessness. Whatever the motive or method, protecting information-carrying signals and documents should be a significant concern of corporations since proprietary information can represent the proverbial keys to the kingdom.

5.2.1 **Energy, waves, and information***

The modern world relies heavily on the electronic transmission of data via wire, fiber, or atmospheric propagation. Audible conversations that exist as mechanical energy propagating through matter also represent an important form of communication. Information in the hands of someone who does not have a "need to know" can be quite revealing and unintended detection and misuse of information so derived can result in embarrassment and/or financial loss.

The conveyance of information is accomplished via the propagation of signal energy that passes through various physical media such as air,

---

*An elegant book on physics and its relation to information is "Electrons, Waves, and Messages" by John R. Pierce (Hanover 1956).

wood, steel, glass, etc. The energy is characterized by repeated cycles of changes in the signal's amplitude. The repeated cycles of energy are called waves, and sensors (e.g., the human eye, the human ear, a microphone) are designed to detect the intensity of this energy. The interaction of these waves with various media can be explained by fundamental principles of physics. These principles play an important role in characterizing the vulnerability to information loss through unauthorized signal detection and misuse of the information contained therein.

What is a wave of energy? We are quite familiar with the general concept by virtue of visits to the beach. Water-borne energy gives us an unforgiving wallop if we try to stand between it and the shore before it crashes on the beach. The world is replete with examples of waves that are often unacknowledged such as the reflected light you are conveniently exploiting to read this page.

Energy in the form of waves that convey information can exist in two distinct physical forms: mechanical and electromagnetic. We will see that each propagates through matter in different ways and are affected by the medium in which they propagate quite differently, although each can be characterized using equivalent mathematical representations. Signal energy and associated transmission of information occur in the form of oscillatory motion where the amplitude of that energy changes in time.

The amplitude of a wave corresponds to its height as measured from its mid-point or zero crossing point to its peak value. The wavelength is the distance over which the wave repeats itself as measured from peak to adjacent peak. The frequency of energy traveling through matter can theoretically vary at any rate, measured in cycles per second or hertz, where a cycle is the time it takes for the wave amplitude to repeat its oscillation. Wavelength and frequency are reciprocally related so shorter wavelength energy waves are by definition higher in frequency. The concepts of wavelength and frequency are depicted in Figure 5.1.

An energy wave oscillating at a rate of a million cycles per second (i.e., 1 MHz) repeats a complete cycle every millionth of a second. Similarly, a wave that oscillates at a rate of a billion cycles per second (i.e., 1 GHz) repeats itself every billionth of a second. Appendix A lists some common scientific prefixes.

All signal energy can be characterized in terms of wave motion that oscillates at a specific frequency or set of frequencies. The electromagnetic energy spectrum in particular extends over an extremely broad range

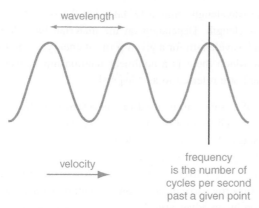

■ **FIGURE 5.1** Wavelength and frequency. *From www.ncbi.nlm.nih.gov/bookshelf/br.fcgi?book=w...*

of frequencies. It is interesting to observe that the portion of the spectrum corresponding to the set of electromagnetic frequencies humans can visually detect (i.e., visible light) represents a small fraction of the total electromagnetic spectrum as shown in Figure 5.2.

The concepts of wavelength and frequency are important in examining the vulnerability to information loss. For signal energy propagating through air, the frequency, f, and wavelength, w, are related through the expression wf = c, where c = the velocity of the wave. If one knows the velocity and frequency of a wave of energy in a particular medium it is possible to

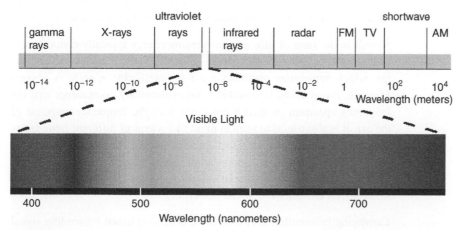

■ **FIGURE 5.2** Electromagnetic spectrum. *From http://www.dnr.sc.gov/ael/personals/pjpb/lecture/lecture.html.*

determine the wavelength. Similarly, knowing the velocity and frequency yields the wavelength. Depending on the material, the relation between frequency and wavelength for a given form of energy is not always linear. Materials for which there is a nonlinear relationship between frequency and wavelength are referred to as "dispersive."

The concept of a wave can be used to characterize all the energy that we hear and see as well as a lot of energy that we cannot sense at all without special equipment. We exist in a world in which we are continually awash in a bath of energy that goes mostly unnoticed. A key point is that electromagnetic waves differ from mechanical waves in fundamental ways, but both forms of energy carry information that can be vulnerable to unauthorized detection by an adversary.

Electromagnetic waves oscillate in a plane perpendicular to the direction of motion and are therefore referred to as transverse waves. In contrast, mechanical waves propagate by displacing the media in which they travel in the direction of motion and are referred to as longitudinal waves. The distinction between electromagnetic and mechanical waves is important in assessing vulnerability to the detection of information-carrying signals.

The human ear responds to mechanical vibrations in the form of pressure variations of air ranging from about 20 Hz (cycles/s) up to a maximum of about 20,000 Hz (i.e., 20 kHz) of acoustic vibrations. This explains being oblivious to ultrasound in contrast with man's best friend. Dogs can hear acoustic vibrations up to about 60 kHz. As noted above, the human eye has a quite narrow frequency response when compared to the entire electromagnetic spectrum. This explains why special detectors must be used to augment human vision in the infrared portion of the spectrum.

Signals in the form of information-carrying energy typically contain more than one frequency. Acoustic energy launched by the human vocal apparatus, which sometimes manifests itself as intelligible speech, is a good example. As an aside, it is probably no evolutionary coincidence that the frequency spectrum of speech coincides with the frequency response of the ear. It is also useful to keep in mind that waves of different frequencies will travel at different velocities depending on the medium in which they are propagating. This fact also has important consequences for the conveyance of information and associated issues of vulnerability to information loss through unauthorized signal detection.

Conveying information over a distance is accomplished by sending signal energy from a point of transmission to a point or indeed multiple points of reception. The information to be conveyed is encoded into the transmitted signal and then decoded upon reception. The encoding is known as modulation, which entails changing some characteristic of the transmitted wave.

For example, most commercial radio stations broadcast in either the AM or FM bands. AM and FM are acronyms for amplitude modulation and frequency modulation, respectively. Information such as speech or music is transmitted to the listener by modulating or changing either the amplitude or frequency of a carrier wave of energy radiated through the atmosphere. The carrier signal is then detected by AM or FM antennae in the signal's path of propagation, demodulated, and amplified by radio receivers. This signal is then used to drive a set of audio speakers, which generate a reasonably faithful reproduction of the information contained in the original broadcast.

As noted earlier, it is a simple but important fact in analyzing the vulnerability component of risk that the energy of information-carrying signals exists in only two physical forms: mechanical or electromagnetic. Analyzing the vulnerability component of risk for the threat of information loss through unauthorized signal detection and other physically based threats is dependent on knowing the form of signal energy used to transmit the information. The behavior of signal energy in transit through a particular medium could be very different in each case. For that reason I will examine each of these separately.

5.2.2 Introduction to acoustic energy and audible information

Acoustic energy in the form of audible vibrations called sound is an example of mechanical energy. It is also an important form of energy with respect to security risk due to the potential for information loss from the unauthorized detection of speech. In simple terms, the physics of sound propagation initially involves a person (hereafter referred somewhat disrespectfully as "the source") opening her mouth to speak. In so doing, she vibrates a set of vocal chords that in turn mechanically excite air molecules in the throat. These molecules shake nearest neighbor air molecules until the energy of vibration is dissipated to the point of extinction through friction and heat. By changing the shape of the mouth as well as the vibration of the vocal chords, a person can modulate the tone and generate words and inflection.

Let's examine the acoustic waves generated by speech after the initial vibration by the vocal chords. The resulting acoustic energy shakes nearest neighbor air molecules in a continuous chain of vibrations that propagate away from the source. Sound is a form of mechanical energy representing

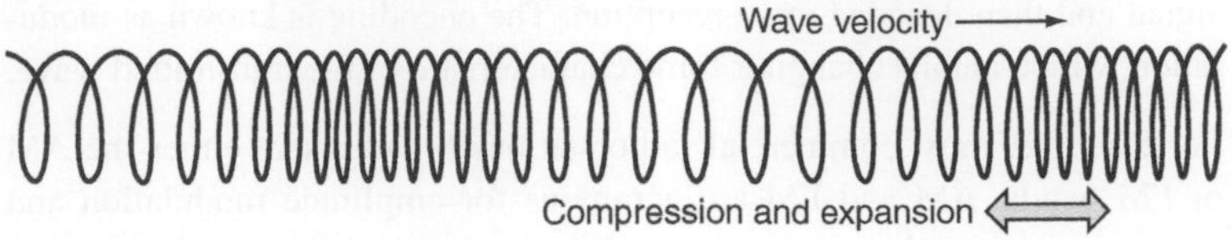

Wave velocity ⟶

Compression and expansion ⟺

■ **FIGURE 5.3** Longitudinal wave dynamics.

pressure fluctuations in the medium in which it is propagating. These waves are called longitudinal waves and the changing amplitude of the wave is in the same direction as its propagation. This process is nicely represented by the motion of a slinky toy as depicted in Figure 5.3.

Upon vibrating the initial set of air molecules, a continuum of molecular vibrations cascades in ever-increasing radii from the source. Each vibrating molecule touches some number of neighboring molecules setting in motion an expanding sphere of vibrations forming areas of compression and rarefaction (i.e., expansion) characterizing pressure fluctuations in all directions.

What happens as the acoustic energy wave radiates away from the person's mouth and vocal tract? As noted above, this wave of sound energy propagates through the air by shaking nearest neighbor molecules of air. This process does not continue unabated forever, otherwise you would have the dubious pleasure of listening to every sound created everywhere in the world.

Fortunately the sound intensity is reduced by a factor proportional to the increasing surface area of an imaginary sphere. If you are worried about your next-door neighbor overhearing arguments between you and your spouse, or unauthorized persons overhearing discussions between you and your business partner while on a train, it would be useful to understand the sound intensity as a function of distance from the sound source.

Recall the concept of point sources of radiation and refer once more to Figure 2.8. The total power of a signal in the form of an acoustic energy wave does not change with distance from the source, but its intensity or power per unit area does. The signal intensity becomes increasingly dilute as the signal energy moves through the medium in which it travels and gets further from the source.

Point sources, i.e., sources that are small in comparison to the wavelength of the energy they radiate, transmit energy *isotropically*. This means the power per unit area is the same in all directions at a given radius from the source. There are also many examples of non-point source radiation sources in security and elsewhere in the physical world. These are collectively referred to as spatially extended sources. However, a number of important security problems can be modeled using point sources. It is worth keeping in mind the image of the expanding sphere as a model for the signal intensity as a function of distance from the source. It is the intensity of this acoustic wave that the incredibly sensitive human ear detects.

Since intensity scales inversely with the square of distance for point sources of energy, this relationship is by definition nonlinear. As discussed in Chapter 2, a linear dependence between two variables is where changes in one variable result in proportional changes to the other. This is not the case for the relationship between signal intensity and distance for point sources: doubling the distance reduces the intensity by 4, quadrupling the distance reduces the intensity by 16, etc.

Before continuing we should describe the concept of intensity in more detail. From a security/vulnerability perspective, knowing the total energy associated with a radiated signal is sometimes less useful than knowing the signal power. Signal power is defined as the time rate of change of radiated energy, and this quantity is quite relevant to the problem of signal detection. A 20 W lightbulb radiates the same amount of energy in 5 hours as a 100 W bulb in one hour. But the total energy radiated by the bulb is of little consequence when we want to read a book. It is the energy radiated per unit time that is important since instantaneous detection of the reflected light by the eye is the critical factor in reading.

The intensity of a signal can be defined as the signal power (i.e., energy/ time) as measured over a specified area. It is therefore the time rate at which energy is deposited on or through a surface of a given area. When a sensor such as the human ear detects sound energy, it is the intensity of the acoustic pressure wave that is detected by the eardrum. So a change in intensity can result from a change in the signal power and/or by a change in the physical area impacted by that energy. For a fixed signal power, the smaller the area impacted by the signal energy the greater the signal intensity.

5.2.3 Transmission of audible information and vulnerability to conversation-level overhears

Conference rooms and offices are rarely built with a concern for audible information loss through unauthorized detection of speech. Yet sensitive information is often discussed in areas contiguous with physically uncontrolled space. In addition, business people routinely discuss sensitive information in public places and are oblivious to those around them. In the case of discussions in public, a little education on vulnerability could go a long way to reduce the risk of this type of unauthorized signal detection. It is important to at least understand the limits of vulnerability in these and other common security scenarios.

We can actually estimate the theoretical vulnerability to unauthorized overhears of conversations. Let's suppose two senior executives of the American Widget Corporation (AWC) are conducting a sensitive conversation in the (ultra) quiet car of an Amtrak train. The intensity of the sound energy in their verbal exchange is at the subdued conversation level. There is a man wearing a T-shirt with the logo of a fierce competitor, the World Gadget Corporation, sitting across the car from the two executives.

Some admittedly very unrealistic physical conditions will be assumed. First, the train is acoustically isolated from the outside world so there is no external audio interference. No one else in the car is speaking and no outside acoustic energy enters the car so there is no background interference in the relevant signal bandwidth. In addition, there is an unobstructed air path between the two company executives and the individual from World Gadget Corp and no energy is absorbed between source and adversary. Finally, we assume that the interior of the car reflects no acoustic energy, which is also completely ludicrous in the real world. Accepting these conditions, what is the American Widget Company executives' vulnerability to having their conversation overheard by the overly inquisitive guy from World Gadget Corp?

Let's assume the power, P, of the executives' conversation-level speech is 10^{-6} W. From the inverse square law, at a distance, $r = 3$ m the intensity from this point source is given by $P/4\pi(3)^2$. This is equivalent to an intensity of 10^{-6} W/(113)m$^2 = 8.8 \times 10^{-9}$ W/m^2.

The minimum threshold of human hearing at 1 kHz has been experimentally determined to be 10^{-12} W/m^2 with pressure variations equal to 2.0×10^{-5} N/m^2 or 2.0×10^{-10} that of normal atmospheric pressure. These figures are a testament to the incredible sensitivity of the human ear. Below that intensity the human ear is incapable of responding so there is no vulnerability to being overheard by a third party. This can be a useful security metric for some scenarios.

So the intensity of conversational sound energy that is "available" to the nosy man from World Gadget Corporation is $8.8 \times 10^{-9}/10^{-12}$ or 8800 times above the minimum sensitivity threshold of the human ear. The vulnerability component of risk can be expressed perfectly well as 8800 times this "zero vulnerability" condition. Zero vulnerability in this scenario occurs when the signal intensity equals the minimum threshold of sound detection by the human ear. This scenario does not bode well for the American Widget Corporation executives who happen to be discussing the company's five-year business strategy. It is clear that American Widget Corp needs a good security director.

What distance of separation between source and listener would be required to achieve zero vulnerability to unauthorized conversation-level overhears? Remember that I am conjuring up a very contrived scenario devoid of ambient noise, signal absorption, and signal reflection. The acoustic power radiated by each American Widget Corp executive when speaking is 10^{-6} W and there are no other audible noise sources of energy. Our goal is to figure out the distance at which the intensity of the executives' speech equals the minimum threshold of human hearing. This is equivalent to solving for the distance, r, in the expression for the sound intensity:

$$10^{-6}\text{W}/(4\pi\,r^2) = 10^{-12}\text{W}/\text{m}^2 \tag{5.1}$$

We find that this distance is an amazing 282 m (an American football field is about 90.9 m). Therefore, a minimum physical separation of 282 m or greater is necessary to achieve zero vulnerability to the threat of being overheard by the man from World Gadget Corp. Invulnerability exacts a price and in this idealized world that price is significant physical separation.

We know from experience that this result is nonsense. Why? A major simplification in this analysis has been to neglect the presence of noise, a ubiquitous phenomenon in the real world. Let's get real and introduce background noise into the scenario and see how it affects the required separation to achieve a zero vulnerability condition.

This time the nosy neighbor must contend with an acoustic background intensity that directly interferes with the target conversation intensity. Let this acoustic background intensity be 10^{-8} W/m^2. This then becomes our new threshold for audibility since any acoustic signal in the train that falls below this intensity level will be swamped by the competing background intensity of audible signals.

Using 10^{-8} W/m^2 as the threshold intensity and again solving for the separation distance, r, we find that the distance is now approximately 3 m. This is a far more reasonable answer and tells us that for distances greater than about 3 m between speaker and listener there is a zero vulnerability to signal detection by an overly inquisitive passenger. This is obviously a strikingly different result from that of a noiseless environment but serves as a good lesson on the importance of considering ambient noise for physical processes related to security.

In the interest of full disclosure I have also neglected the effects of ambient temperature and pressure on acoustic energy. The effects of these conditions are cumulative and can be significant over appreciable distances. Such distances are usually in excess of those considered in assessing vulnerability to the unauthorized detection of conversation-level overhears. However, to provide a realistic assessment of physical security risk one must *always* account for the intensity of background noise in the relevant frequency regime.

5.2.4 Audible information and the effects of intervening structures

We have thus far neglected the effects of intervening structures in examining their effect on the transmission of audible information. Let's examine some of the physical effects of these objects that can influence estimates of vulnerability to information loss through unauthorized signal detection.

We have just learned that sound energy is conducted by vibrating a successive chain of nearest neighbor molecules within the material it is propagating. Possibly contrary to intuition, dense materials conduct sound better than less dense ones. Refer to Appendix C for the speed of sound in various materials. For example, the speed of sound in wood is roughly 11.6 times that of air so sound energy conducted through wood travels a considerable distance before being attenuated through friction.

Anyone sharing the subway platform with me each morning experiences this phenomenon firsthand as we hear the sound of the approaching train conducted by the steel rails in advance of the acoustic energy propagating through air. The speed of sound in steel is 17.8 times the speed of sound in air.

Returning to the effect of intervening structures, although it is true that a considerable fraction of the incident acoustic energy is reflected at an air–wood interface, the speed of sound in wood relative to that of air suggests it would be a mistake to think that this small fraction of sound energy is not conducted by the wood structure. This mistake would be compounded by assuming that the sound energy conducted by the wood could not be detected in other parts of the building.

We know from personal experience that the intensity of sound decreases from one side of a dense barrier to the other. Mechanical forms of energy like sound are sensitive to changes in the *density* of the material in which they propagate. So when sound energy in the form of vibrating air molecules meets the denser-than-air wood barrier, much of the propagating energy is reflected. Some of the energy is actually absorbed by the wood and the remaining energy makes it across to the other side as it resumes its excitation of nearest neighbor air molecules. Acoustic engineers speak of the "acoustic impedance mismatch" between materials and great attention is paid to this phenomenon in designing acoustic structures such as concert halls.

In the case of an acoustic wave, a ballpark estimate of the fraction of signal reflected back to a source by an intervening barrier can be made by first forming the product of the velocity of the sound wave with the density of the material through which it travels. This is the definition of acoustic impedance. We then form a ratio of the difference of this product for the two materials divided by the sum of this product of the two materials and square the result. In other words, we form the ratio of the difference in acoustic impedance of two dissimilar materials divided by the sum of their acoustic impedances and square the result to obtain the fraction of the reflected signal that results when an acoustic energy wave traveling in one medium suddenly confronts a completely different medium.

If d_1 equals the density of air, d_2 equals the density of a wood intervening barrier, v_1 equals the velocity of sound in air, and v_2 is the velocity of sound in wood, the fraction of the total power that is reflected is given by

$$R^2 = ((d_1v_1 - d_2v_2)/(d_1v_1 + d_2v_2))^2 \qquad (5.2)$$

The remaining energy enters the wood and is either absorbed or transmitted to the other side of the wall. The sum of the fractions of transmitted, reflected, and absorbed sound energy equals the total power in the signal, or $T^2 + R^2 + A^2 = 1$.

One must be careful with this type of analysis for rooms that are highly reverberant or "live," i.e., those lacking absorbing materials. In acoustically live rooms, even though the wall may transmit only 1% of the acoustic energy, 99% of the reflected energy continues to bounce off the walls until some of it leaks into the adjacent room. Even a massively heavy wall will not kill all the transmitted energy in a highly reverberant room.

The theoretical vulnerability to information loss through unauthorized (and unaided) detection of acoustic energy in an open space was previously calculated in the scenario of the two American Widget Corp executives on a train. Security professionals interested in protecting sensitive information should also be concerned with audible information loss resulting from conversations within rooms. This can be a complicated technical problem, but some simple models are provided for ballpark estimates of the vulnerability component of risk.

For example, we could compare the intensity of acoustic power incident on a wall of a room compared to the power transmitted across that wall into the next room. This would be a simple metric for the vulnerability to information loss through unauthorized audio signal detection. The transmission loss (TL) = $10\log I_{inc}/I_{trans}$. The logarithm is used in this case because the ratio I_{inc}/I_{trans} is a large number, and as we learned in Chapter 2, taking the logarithm makes the expression for the ratio smaller. Transmission loss is given in units of decibels, which were introduced in Chapter 2.

The wall will vibrate in reaction to the acoustic energy of the successive, chain-like vibration of air molecules stimulated by the speaker. The frequency response varies across the audible spectrum and depends on the mass of the wall.

At low frequencies, the wall is susceptible to so-called resonances where it absorbs more energy and vibrates preferentially at that frequency. Resonance is an important physical phenomenon that occurs in both mechanical and electromagnetic systems. In general, resonance effects occur over a well-prescribed bandwidth and are directly related to specific physical properties of the system relative to the excitation energy.

At very low frequencies, the wall can be modeled as a mass attached to a spring in response to an applied force, as depicted in Figure 5.4. In general, the mass-spring model is characterized by the expression $F = -kx$ (Hooke's Law), where F is the applied force, m is the mass, k is the spring constant denoting stiffness or opposition to the applied force, and x is the displacement of the mass. For undamped motion, the angular frequency of vibration is given by $\sqrt{(k/m)}$.

This is an extremely important model in physics and is used to describe many physical processes. A similar mass-spring model will be invoked to describe the response of windows to explosive forces. In the case of the wall, the spring describes the restoring force that the wall exerts in opposition to the sound pressure attempting to bend the wall.

For low mass walls, higher frequencies might also exhibit resonance effects at the so-called coincidence frequency where the wavelength of induced ripples along the wall surface matches the incident acoustic energy wavelength. In this case resonance occurs in the 1000–4000 Hz

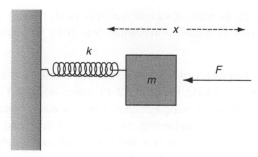

■ **FIGURE 5.4** Mass and spring indicating harmonic motion.

range resulting in a more efficient transfer of acoustic energy across the wall boundary with the attendant risk of unauthorized detection of signal energy. Note that these frequencies are roughly in the middle of the frequency band for human speech.

At very low frequencies or those below 100 Hz, the spring force dominates the wall's response, and the transmission loss is given approximately as

$$\text{Transmission Loss (TL)} = C - 20 \log(fR^4/Yh^3)(\text{dB}) \qquad (5.3)$$

where C is a constant, f is the sound frequency, R is a typical transverse dimension between wall supports, h is the thickness of the wall, and Y is the wall stiffness (the Young's modulus. . .no relation to the author) that corresponds to k in Figure 5.4. In this case the effect of wall stiffness decreases as the sound frequency is increased.

At higher frequencies between 100 Hz and 1 kHz, the transmission loss from one room to another separated by a wall is given by the following expression:

$$\text{Transmission Loss (TL)} \leq 20 \log(f\sigma) - 48(\text{dB}) \qquad (5.4)$$

In this expression, f is the frequency of the acoustic wave and σ is the mass density of the wall in units of kg/m^2. In this frequency regime, each doubling of sound frequency increases the transmission loss by a factor of two (note that each doubling of frequency is referred to as an octave, a term that is familiar to anyone who has studied music). It is clear from this expression that the two physical parameters of importance are the density of the wall material and the frequency of the incident sound energy.

Since audible acoustic energy for humans ranges from a few hertz to nearly 20 kHz, we must bear this in mind when we evaluate the transmission loss, TL. The TL for walls of various materials yields ballpark estimates of the vulnerability to audible information loss from unauthorized signal detection.

For example, half-inch gypsum board has a mass density σ of 10 kg/m^2. At a frequency f of 1 KHz, the TL is 32 dB, which is a useful security risk metric if this material is being considered as an acoustic barrier.[1] The 32 dB of transmission loss implies that the intensity of the sound in the room where the speaker is located is a factor of 1584 times higher than the intensity on the other side of the wall at that frequency.

On its face this seems like a lot of attenuation, but what does this really say about the vulnerability to information loss with respect to the threat of conversation-level overhears? Certainly we need to account for the human ear's tremendous sensitivity to know whether a factor of 1584 in attenuation is sufficient to reduce the intensity of acoustic energy such that the vulnerability component of risk to unauthorized signal detection is effectively zero. Equivalently we need to determine if the energy intensity is below the detection threshold of human hearing.

If the conversation level sound intensity in a room separated by a wall from an adjacent room is 10^{-5} W/m^2, then the transmission loss across the wall is 10^{-5} W/m^2 minus 32 decibels of attenuation as calculated above. This is equivalent to 10^{-5} W/m^2/1584 = 6.3 $\times$ 10^{-9} W/m^2. If the background intensity is 10^{-8} W/m^2 (i.e., library-level background noise) on the nosy listener's side of the wall, then the vulnerability to audible information loss is still a factor of 1.6 above zero vulnerability *at 1 KHz*.

This is a ballpark approximation as site-specific conditions can vary considerably to include the existence of hidden pathways of acoustic transmission across the wall boundary. Also, there is nothing to prevent the nosy person in the next room from directly coupling a device like his ear to the wall and exploiting the structure-borne signal energy. Worse, these same people could poke a tiny hole through the wall. In these cases this simple calculation goes out the window. As noted previously, lower frequency components of speech will be less attenuated by the gypsum board barrier.

What happens to the vulnerability to unauthorized signal detection if a 4-inch brick wall separates two rooms? Brick has a mass density σ of 200 kg/m^2, which yields a TL of 58 dB or a factor of almost 631,000 (i.e., $10^{5.8}$) at 1 KHz.[2] The attenuation would double for every doubling of the acoustic frequency in the so-called mass-controlled region of the frequency spectrum (i.e., 100 to 1000 Hz).

There is also a doubling of the sound attenuation for each doubling of the wall mass in the 100 to 1000 Hz frequency range. However, one could imagine that this method of risk mitigation could be quite impractical if the weight of the wall is of any concern. A more practical solution is given in Chapter 6. A table summarizing the frequency dependence of acoustic attenuation by walls/barriers is listed in Table 5.1.

Unfortunately even this situation is somewhat idealized. We assumed that the transmitted speech energy left the originating room and was partially reflected by the wall while the transmitted portion decayed in intensity after entering the room next door. In fact, and as was pointed out previously in noting the effects of reverberation, it is likely that the reflected energy continues to bounce around the room from which it originates contributing to the intensity of the transmitted signal. In that case the vulnerability to information loss is increased. The TL figure should be treated as the most optimistic view when evaluating vulnerability to conversation-level overhears.

Finally, all this detail about the transmission of sound through walls to determine vulnerability to information loss becomes relatively inconsequential if there is a more convenient path (i.e., smaller acoustic impedance mismatch) for the sound to travel. Such paths might include a ventilation duct or the space above a suspended ceiling. A ballpark estimate on the sound reduction expected between rooms with a common suspended ceiling of ¾" fiber tile (density $\sigma = 6 \text{ kg/m}^2$) with a 60 cm utility space above will at best result in a sound intensity reduction between a factor of 300 and 1000.[3]

This may sound quite effective, but if we again assume conversation-level acoustic intensity in a room corresponds to about 10^{-6} W/m^2 and the background intensity in the room next door is 10^{-9} W/m^2, even a factor of 1000 drop in sound intensity would yield acoustic levels just at the theoretical limit of audibility. Since the numbers listed in Appendix B for

Table 5.1 Frequency Dependence of Wall/Barrier Acoustic Attenuation

	Wall/Barrier Acoustic Attenuation Factors
<100 Hz	Stiffness is predominant in attenuation; resonances exist
100–1000 Hz	Doubling the mass or the sound frequency doubles the attenuation
1000–4000 Hz	Resonances @ coincidence angles

sound intensities are estimates, and on-site physical conditions can vary considerably, this calculation of sound reduction should be considered imprecise at best. There may still be an appreciable vulnerability to audible information loss due to unauthorized signal detection for this scenario, but at least your eyes (if not your ears) have been opened to the magnitude of risk.

5.2.5 Introduction to electromagnetic energy and vulnerability to signal detection

Electromagnetic energy waves carry information and are produced by radio transmitters, mobile phones, laser communicating devices, Blackberries, wireless headsets, Bluetooth technology, etc. It is worth taking a moment to ponder the continuous bath of man-made electromagnetic energy to which we are exposed during every moment of every day.

As its name implies, electromagnetic energy consists of electric and magnetic components. These oscillate out of phase with each other and are mutually perpendicular in their respective planes of oscillation. In contrast with mechanical waves, the direction of electromagnetic wave propagation is perpendicular to the planes of electric and magnetic field oscillation. For this reason electromagnetic waves are known as transverse waves.

Although all electromagnetic energy has both an electric and magnetic field component, detectors such as conventional antennae respond better to the electric field. In 1864, a British physicist named James Clerk Maxwell formulated four equations that are at the heart of all electromagnetic phenomena. These are now commonly known as Maxwell's equations. Needless to say there are an almost uncountable number of references that discuss these expressions, and I won't say much more about them except to point out basic concepts that are relevant to security.

Maxwell's equations specify that the time rate of change of the flux of electric and magnetic fields "flowing" through a region due to static charges and currents, will produce magnetic and electric fields, respectively. From a security perspective, the resultant fields are subject to unauthorized detection and potential information compromise.

Because of the physics of electromagnetic versus mechanical energy propagation, the former typically travels through air farther than the latter, which has rather obvious implications with respect to security risk. In this chapter key risk factors for electromagnetic signals are identified in order to assess the vulnerability component of the risk of unauthorized signal detection.

We observed in our previous discussion on audible information that mechanical energy causes air molecules to shake producing areas of compression and rarefaction as the pressure wave radiates away from the speaker. These vibrating molecules arouse sympathetic vibrations from their neighbors in a continuous manner until the energy dissipates due to friction. If there is no intervening medium to vibrate then there is no possibility of sound propagation. The best insulator of sound energy would be a vacuum, which by definition is totally devoid of atoms and molecules.

In stark contrast, electromagnetic energy needs no physical medium to support propagation and travels quite contentedly in a vacuum. For example, electromagnetic signals from earth are sent to satellites and space probes all the time. That is not to say that intervening materials do not affect electromagnetic wave propagation. We routinely witness such effects every day.

To illustrate vulnerability to unauthorized detection of electromagnetic energy, let's say that the International Signal Corporation (ISC) wireless LAN is radiating unencrypted packets and the facility is located about 10 m from a building housing ISC's chief competitor, United Telemetry Incorporated (UTI). The director of security at ISC is quite concerned that UTI may be detecting their information-rich signal but the ISC company president, a graduate of a prestigious MBA program, is too cheap to encrypt the signal.

To the amazement and horror of its Board of Directors, ISC has been suspiciously out bid by UTI on a number of sensitive and potentially lucrative deals. The security director suspects a leak, and the justifiably paranoid company must now hire a very expensive security consultant, a graduate of a prestigious engineering program, to investigate.

Acoustic energy emanating from people is not the only example of point sources of energy. Some radio transmitter antennae, such as those seen on top of high-rise buildings, also radiate energy in all directions. The antenna on top of the Empire State Building is one such example and it is an impressive 206 feet tall. Since it radiates energy isotropically, the intensity can be characterized by the now-familiar spherical pattern.

Not all antennae transmit energy in this manner. For example, the intensity of energy radiating from a dish antenna is quite different than from a dipole. Dish antennae, a familiar site on roofs across the world as they are used for geosynchronous satellite links to receive television signals, receive and transmit energy within slices of space. The power density within that slice is enhanced at the expense of the remaining slices. In this way the antenna demonstrates geometric *gain*, a defining feature of directional antennas.

It is assumed that ISC's commercial LAN transmitter is a point source and is radiating energy isotropically. After all, companies usually want as many of their employees to access the signal as possible, so it is important to transmit over a wide area to optimize coverage. We need to know the output power of the transmitter in question as well as the distance between the transmitter and a given location.

In the interest of complete disclosure, this estimate is fudged a bit since I am neglecting the effects of intervening material such as walls, doors, etc. We would expect such materials to absorb the LAN signal depending on the transmission frequency as well as for the signal to be reflected off some surfaces, especially metallic ones. It is always good to bear in mind that reflections can work just as easily in your favor as it can for an adversary. The effect of intervening material on radio frequency signals will be discussed in Section 5.2.6.

Since we have a point source radiating energy, the inverse square law applies and the signal intensity scales as $1/r^2$ where r is the distance between source and the point of reception. We can immediately calculate the intensity as a function of distance in typical fashion. However, the signal intensity is only part of the story. We must also speculate on the capture area of UTI's antenna to calculate the signal power at the point of detection. Unfortunately we do not know the technical characteristics of the UTI antenna. So our estimate in this case will be just that . . . . an estimate based on reasonable system parameters and scenario-driven conditions.

Specifically, we must multiply the estimated radiated intensity at the receive site times the capture area of the UTI antenna. Two obvious choices for an antenna type are the dish and simple dipole. Both are familiar to those who own a satellite TV system and/or an FM stereo. All antennae are not created equal, and their signal gain varies according to the physical design. Since ISC's LAN is radiating isotropically, the signal intensity at a given distance from the source is the same in any direction. UTI's receive antenna can only capture a fraction of the total transmitted radiation. For a dish antenna this fraction of energy is proportional to the area of the dish itself (i.e., πR^2, where R is the radius of the dish).

The dipole is another reasonable candidate for a receive antenna in this case. The effective signal capture area of a dipole antenna is approximately equal to $\lambda^2/4\pi$, where λ is the signal wavelength. This expression tells us that the dipole antenna capture area is greater at lower frequencies, or equivalently, longer wavelengths. UTI also has the benefit of knowing the transmit frequency of the ISC system as it probably conforms to the standard 802.11 LAN protocol.

We must also estimate the noise power produced by the predominant source of noise at the transmit frequency, since it is ultimately the ratio of signal power to noise power that determines vulnerability to unauthorized signal detection. Electromagnetic noise is generated from a variety of natural and artificial sources, including electronic devices, sunspots, lightning, etc. In addition, the noise power varies significantly across the electromagnetic spectrum. A true noise signal is a random, nonrepetitive phenomenon, and the sources of noise in the electromagnetic regime are quite varied. The reader is referred to "Radio Noise. Recommendation ITU-R" P.372-7, International Telecommunication Union, Geneva, 2001, for comprehensive data on sources of ambient radio frequency noise.

Figure 5.5 depicts levels of measured ambient electromagnetic noise as a function of frequency generated by man-made and natural sources.[4] Median values of man-made radio noise power (solid lines) are expressed in terms of F_{am} (dB above the thermal noise limit or kT_0 where k is Boltzmann's constant, T_0 = absolute temperature at room temperature = 293 K, and B is signal bandwidth). Atmospheric noise (dashed lines) and cosmic background noise (dotted line) are shown for comparison with man-made noise.

The quantity kT_0B represents a fundamental noise limit that sets the minimum threshold for signal detection. This limit is determined by the motion of atoms in the components of the receive equipment such as the antenna and is directly related to their temperature. No amount of fancy signal processing can overcome this limit imposed by Mother Nature. Other typical sources of competing electromagnetic noise are actually much higher in amplitude than electronic thermal noise and this makes life more difficult for an adversary.

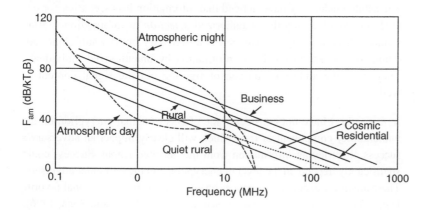

■ **FIGURE 5.5** Frequency and amplitude of ambient noise sources relative to kT_0B.

The important point is that an assessment of the vulnerability to the detection of signal energy is always made in relation to the ambient noise power. If the S/N ratio is less than ten, a commercial adversary's success at reading a signal of interest is problematic. We can again express the vulnerability to signal detection as some factor above zero risk where zero risk is defined as a S/N ratio of unity.

When we perform the vulnerability to detection by UTI we see that there is roughly a 39 dB S/N ratio at 10 m (see the section in this chapter titled Security Metric: The Vulnerability to Unauthorized Radio Frequency Signal Detection). From ISC's perspective this is not good. The security consultant for ISC cannot prove UTI is detecting the signal and attempting to retrieve its information content, but the consultant must highlight the magnitude of the vulnerability component of risk.

One must try and appreciate what can and cannot be controlled in this scenario. There is nothing that can be done about the ambient noise level. Urban environments typically experience higher levels of man-made noise than rural locations and the noise power at certain frequencies can be considerable. The ambient noise needs to be high to decrease the vulnerability to interception, but not so high that the system becomes useless to the intended recipients.

This is where the inverse square law for intensity and distance works against an adversary. In theory ISC could turn down the transmitted power of its LAN signal, but then it runs the risk of its employees not being able to detect the signal. It could physically move its building away from the competition contributing to the $1/r^2$ signal loss. This would be effective but would most likely not be considered very practical to the people controlling the budget.

The astute reader may have noticed that information loss scenarios considered here involve only the unauthorized *detection* of information-bearing signals. Of course, signal detection is a necessary but not sufficient condition for success in obtaining information by unauthorized individuals. They have the additional burden of figuring out the signal content as well.

In that vein, I have not considered the effects of signal encryption on the vulnerability component of risk, which will clearly impact an adversary's success at retrieving information from the detected signal. Success could significantly depend on the level of encryption and the particular adversary. There are many excellent references on encryption and my personal favorite is *Applied Cryptography* by Bruce Schneier (John Wiley and Sons, 1996).

With respect to discerning unencrypted but nonetheless complex signal processing schemes from commercial communication devices amid noisy background levels, I refer the reader to references in the public domain that provide technical details: "Optical Time-Domain Eavesdropping Risks of CRT Displays," Markus G. Kuhn, Proceedings 2002 IEEE Symposium on Security and Privacy, and "Electromagnetic Eavesdropping Risks of Flat-Panel Displays," Markus G. Kuhn, 4th Workshop on Privacy Enhancing Technologies, 26–28 May, Toronto, Canada. Also, "Security Limits for Compromising Emanations," Markus G. Kuhn, CHES Proceedings, 2005, and other papers by Kuhn that can be found at http://www.cl.cam.ac.uk/~mgk25/, provide an excellent analysis of the vulnerability of electronic equipment to unwanted interception.

SECURITY METRIC: THE VULNERABILITY TO UNAUTHORIZED RADIO FREQUENCY SIGNAL DETECTION

We first assume a 2.4 GHz wireless LAN is transmitting 100 mW of power isotropically with a signal bandwidth of 10 MHz. At room temperature, the thermal noise limit for a 10 MHz bandwidth signal is calculated to be 0.04×10^{-12} W. A radio receiver cannot detect a 10 MHz signal of lower power.

At a distance of 10 m, it can be shown (see below) that the signal strength is approximately 0.1 μW (i.e., 10^{-7} W). This means that ratio of the power of the signal to the power of the noise is $(0.1 \times 10^{-6})/10^{-12} = 10^5$ or 100,000 times above the thermal noise limit.

However, the electromagnetic noise generated by the sun at a frequency of 2.4 GHz is roughly 25 dB or 316 times greater than the thermal noise limit, and I assume this to be the predominant noise source at that frequency. Therefore, this reduces the ratio of signal noise power to ambient noise power by $100,000/316 = 316$.

This is the result we seek since we now know that the signal strength from the LAN at 10 m is approximately 316 times above the most powerful source of noise in that frequency regime. An S/N ratio of 25 dB would be quite susceptible to unauthorized detection (10 dB is considered the minimum requirement). Below we include the underlying calculations for determining S/N ratios for an unspecified radiating electromagnetic device:

1. Calculation of power received at a distance of 10 m from the signal source

   P_r = receive power
   P_t = transmit power = 63×10^{-3} W
   λ = signal wavelength = 0.13 m at 2.4 GHz in air
   $\lambda^2/4\pi$ = radio frequency cross section of a dipole antenna
   r = distance from transmitter to receive antenna = 10 m

$$P_r = (P_t/4\pi r^2) \times (\lambda^2/4\pi) \qquad (5.5)$$

So, 63×10^{-3} W $\times$ 0.02 m²/$(1.26 \times 10^3$ m²$) \sim 10^{-6}$ W = P_r at 10 m

(Continued)

SECURITY METRIC: THE VULNERABILITY TO UNAUTHORIZED RADIO FREQUENCY SIGNAL DETECTION—CONT'D

2. Calculation of the thermal noise limit and signal-to-thermal noise ratio for a 10 MHz bandwidth signal at a distance of 10 m from the signal source:

   The thermal energy of atoms, kT_0, at room temperature = 1/40 eV (k = Boltzmann's constant and T_0 = absolute temperature in degrees Kelvin). The thermal noise power is given by kT_0B (i.e., Johnson noise), where B = signal bandwidth = 10 MHz.

   So, kT_0B = 1/40 eV × 10 MHz = 1/40 eV × 1.6 × 10^{-19} J/eV × 10 MHz
   = 0.04×10^{-19} W-s × 10 × 10^6/s = 0.04 × 10^{-12} W

   Signal-to-thermal noise ratio at 10 m is approximately 10^{-6}/(0.04 × 10^{-12}) = 25 × 10^6 or 74 dB

3. Calculation of the maximum theoretical reception distance:

   Assume we require 35 dB of signal above the thermal limit to receive a signal (i.e., 25 dB of solar noise at 2.4 GHz and a 10 dB signal margin). From (2) above, kT_0B for B = 10 MHz = 0.04 × 10^{-12} W. The minimum P_r = 35 dB above kT_0B = 127 × 10^{-12} W

   Solving for r we get

   $$(P_t/4\pi r^2) \times (\lambda^2/4\pi) = 127 \times 10^{-12} \text{ W which yields } r = 316 \text{ m} \qquad (5.6)$$

4. Calculation of signal power and signal-to-noise where solar noise represents the predominant source of noise in this frequency range at a distance of 10 m from the signal source:

   $$P_r = \frac{63 \times 10^{-3} \text{ W}}{(4\pi) \times (10 \text{ m})^2} \times (\lambda^2/4\pi) \qquad (5.7)$$

   $$= 0.1 \times 10^{-6} \text{ W} = 39 \text{ dB or} \sim 8000 \text{ S/N}$$

5.2.6 Electromagnetic energy and the effects of intervening material

The effect of intervening materials on electromagnetic energy is very dependent on the type of intervening material and energy frequency. We know from personal experience that radio frequency energies in the FM band (i.e., 88 to 108 MHz) can pass through typical building construction materials such as wood, wallboard, and even concrete if the latter is not too thick. This is not the case with higher frequency signals (or equivalently when the wavelengths are shorter) that extend into the upper microwave regime. Microwave links can be easily disrupted by rain or fog. We know quite well that many materials are not transparent at optical frequencies. This is the functional basis for using blinds on windows and in part for wearing clothes aside from keeping warm.

The attenuation of both mechanical and electromagnetic forms of energy by intervening material is also discussed in some detail in Chapter 6. Some materials offer significant opportunities for mitigating the risk of information loss through unauthorized signal detection depending on the material and the frequency of the radiating energy. In general it is best to be conservative in such estimates since key parameters in a given scenario cannot often be confirmed. It is also prudent to perform vulnerability estimates with and without the presence of intervening material. This will yield a range of values that helps to bound the problem and formulate an appropriately conservative risk mitigation strategy. In this section I will mention some basic effects of electromagnetic propagation and save some of the important details for the discussion on mitigation in Chapter 6.

The effect of metals on radio frequency signal propagation can be profound. Depending on the frequency of the energy and conductivity of the metal, electromagnetic energy is mostly reflected at the air–metal interface. Metals are great for electromagnetic shielding but can play havoc with signal propagation for the intended audience.

Many materials that are completely opaque at optical wavelengths (i.e., between 0.7 and 0.4 μ) allow propagation without significant attenuation at radio frequencies. The transparency of materials to electromagnetic energy is dependent on the type of material and the frequency of the propagating energy. The net effect of this can be good or bad depending on the signal and who should or should not detect it.

The precise route that electromagnetic signals follow when traveling from point A to point B is often not easily predicted, especially in urban environments. A well-known effect is multipath, where a signal can take many paths to arrive at a location. This can produce sharp peaks and troughs in signal intensity as a function of location and frequency. Also, the effect of multipath can change over short timescales due to the movement of objects in the various propagation paths and a changing electromagnetic spectrum. It is not uncommon to experience significantly varied signal intensities as the energy from multiple propagation paths adds or subtracts coherently.

Multipath is also highly wavelength dependent where the signal intensity can vary considerably over distances comparable to the wavelength of the signal. Anyone who has listened to an FM radio in the car and experienced signal fades has probably noticed that the reception can change dramatically when the car is moved just a few feet. This represents a distance comparable to FM signal wavelengths. One possible remedy is to change stations, which changes the wavelength of the received signal energy. The frustration you experience when grooving to your favorite tune could simply be caused by multipath interference. So do not hit the dashboard in frustration. Merely move the car a few feet or change the station.

When a wave of energy is traveling within medium A and suddenly encounters medium B causing the energy velocity to change, the energy also changes direction. The classic example of this is observing a spoon in a glass of water as the spoon appears to bend at the air–water interface. This effect is known as refraction. The speed of light in water is less than in air so the light rays bend according to the ratio of the index of refraction of each medium. This effect is known as Snell's Law. It is also important to note that the index of refraction is highly frequency dependent. Fiber-optic cables exploit this phenomenon to confine light to the fiber core.

Diffraction is a phenomenon that causes spreading of a transmitted energy wave front and affects both mechanical and electromagnetic signals. If an audio source is broadcasting at an opening in a wall, the sound spreads as it passes through the opening. In acoustics, barriers erected specifically to attenuate sound must contend with diffraction when the wavelength of transmitted energy is roughly equal to the dimensions of the barrier.

Another important phenomenon of signal propagation is scattering. This effect accounts for why the sky is blue. Also known as Rayleigh scattering, energy from the sun is scattered or re-radiated after interacting with atmospheric particles whose diameter is approximately equal to the diameter of the wavelength of visible light. This scattering effect scales as the inverse fourth power of the wavelength, λ (i.e., $1/\lambda^4$). Therefore, the shorter wavelength/higher frequency blue light is preferentially scattered relative to other frequencies in the visible spectrum. Optical communication devices such as lasers can be greatly affected by fog and smoke. Atmospheric contaminants in general cause signal attenuation due to various types of scattering effects.

Accounting for refraction, scattering and diffraction would be important in a detailed analysis of the propagation of acoustic or electromagnetic energy. However, for many estimates of vulnerability to unauthorized signal detection of radio frequency signal energy, these effects can be ignored except when considering scenarios that involve long distances or the signal energy wavelength is comparable to the dimensions of intervening material.

The goal of the preceding discussion has been to sensitize the security professional to some of the physical issues involved in the propagation of acoustic and electromagnetic energy in the real world. Precise descriptions of energy behavior are the realm of the physicist or engineer. But exact calculations are less important than understanding the factors that grossly affect vulnerability to unauthorized signal detection and to develop an appreciation for the magnitude of such factors under operationally realistic conditions.

The vulnerability of wireless LANs, wireless headsets, and other radiating electronic devices as well as conversations overheard from within offices, conference rooms, or in public spaces are common scenarios and can be important in estimating the vulnerability to information loss through unauthorized signal detection.

It is useful for the security professional to develop at least a gut feeling for how signal intensity scales with distance as well as the effects of intervening materials on signal propagation. Proper attention can then be focused on the most critical parameters pursuant to developing an effective risk mitigation strategy.

5.2.7 **Vulnerability to information loss through unauthorized signal detection: A checklist**

Let's summarize the critical issues for analyzing vulnerability to information loss from the unauthorized detection and subsequent misuse of acoustic or electromagnetic signals. The scope of this problem can seem quite complex. However, I offer a checklist that can be applied to any information loss scenario to help focus on the important issues and key parameters:

1. Is the signal energy mechanical or electromagnetic? Sound energy is mechanical and radio frequency/microwave and optical/laser transmissions are electromagnetic.
2. What is the signal carrier frequency and bandwidth? The bandwidth of speech ranges from 20 Hz to 20 KHz but radio frequency transmitter frequencies can vary tremendously depending on the device. The power of the radiated signal is an important factor in estimating vulnerability to unauthorized signal detection that potentially leads to information loss.
3. What are the intervening materials between the signal source and the receive location? Recognize that signals often take multiple indirect routes in arriving at a destination. Understanding the effect of materials as a function of signal frequency will enhance the accuracy of estimates of the vulnerability component of risk.
4. What is the distance to the nearest point where there is no control over the physical access to internal space? This is the maximum distance of signal security (or conversely the minimum distance of vulnerability) since an adversary could be physically present at that location and not much can be done about it. If signal sources are point sources with energy propagating equally in all directions, the intensity of

information-carrying signals scales as the inverse-square with distance (i.e., $1/r^2$ where r represents distance) in open environments.

5. What is the power of the dominant noise or interfering source in the frequency range of concern? If the transmitted signal power is below the power of the dominant source of noise at the minimum distance of vulnerability at that signal frequency and bandwidth then there is zero vulnerability to information loss from that signal at that distance (and beyond). It is important to evaluate the noise power across the entire signal bandwidth.

Estimates regarding these five questions should yield significant insight into the vulnerability to information loss through unauthorized signal detection. The emphasis here is on insight rather than on a precise assessment. If an estimate of risk is anywhere close to the limit of signal detection for a particular scenario, it is safe to assume that a vulnerability to unauthorized signal detection exists.

5.3 VULNERABILITY TO EXPLOSIVE THREATS

5.3.1 Explosive parameters

Explosive threats are a significant problem in an era of increasing concern over terrorism. Conventional explosives are relatively easy to obtain and there are numerous examples of damage caused by relatively small amounts of explosive material. Furthermore, vehicles in particular offer opportunities for the concealment and delivery of large explosive payloads.

When an explosion first occurs it creates a shock wave associated with the initial blast that creates an overpressure condition (i.e., a pressure greater than the atmospheric pressure at sea level = 14.7 lb/in.). Depending on how this shock wave interacts with the building structure and/or its individual components (e.g., windows), which in turn is highly dependent on their natural period of vibration relative to the duration of the positive phase of the blast pressure, it can potentially cause wall collapse and/or window breakage.

In addition to the overpressure, the so-called explosive impulse plays an important role in the damage caused by explosive detonations. The impulse is the time integral of the wave pressure. To get a feeling for impulse on a visceral level, consider the difference in the damage to your body if you were subjected to a quick punch versus the effect of that punch applied over an extended period of time. The effects of the latter would be much worse on your anatomy. Figure 5.6 shows the time history of overpressure following detonation.

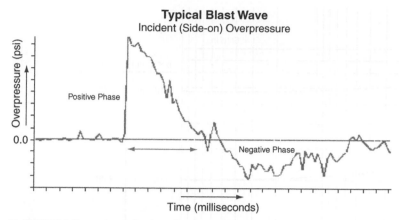

Typical Blast Wave
Incident (Side-on) Overpressure

■ **FIGURE 5.6** Time evolution of explosive blast waves *(Provided by Solutia Inc.)*

There are two key scenario-dependent parameters that significantly affect the vulnerability to explosive incidents: the distance between the explosive source and the target and the explosive payload. These parameters directly influence the magnitude of the explosive-induced overpressure and impulse. It is a combination of the overpressure and impulse that causes the devastating results to buildings when subjected to the force of an explosion. This is demonstrated in Figure 5.7 where a design chart for panels of tempered glass cited in the US Department of Defense TM15-1300 shows the maximum peak pressure that can be sustained as the duration of the pressure is varied.[5]

It is therefore a logical question to ask how pressure and impulse scale with both distance and explosive payload. Surprisingly, there is not complete agreement on the answer to this question. In the technical literature, four distinct scaling laws have been cited for pressure and two for impulse. In one commonly cited work[6] pressure p, and impulse i, scale with payload m and distance r as follows:

$$p(\text{Megapascals}) = 0.085(m^{1/3}/r) + 0.3(m^{2/3}/r^2) + 0.8(m/r^3)$$
$$i(\text{Pascal} - \text{sec}) = 200(m^{2/3}/r) \tag{5.8}$$

Note that impulse scales inversely with distance (i.e., $1/r$) and the dominant pressure term scales inversely with distance cubed ($1/r^3$). Therefore increasing the distance between explosive source and target is clearly beneficial in reducing the vulnerability component of risk mostly because of the very nonlinear scaling of pressure with distance.

After the explosion, the shock wave expands outward and enters the building. It simultaneously pushes upward and downward on the floors. Floor

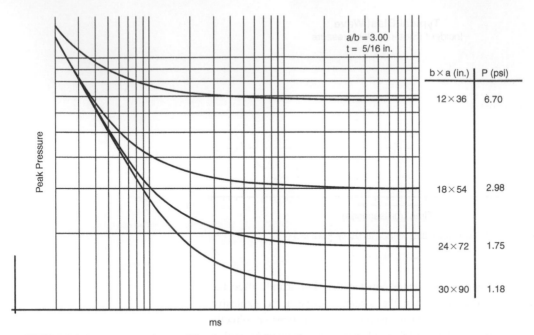

b × a (in.)	P (psi)
12 × 36	6.70
18 × 54	2.98
24 × 72	1.75
30 × 90	1.18

a/b = 3.00
t = 5/16 in.

Peak Pressure

ms

■ **FIGURE 5.7** Peak pressure versus duration of blast for tempered glass panels.

failure is common in these types of explosive incidents due to the large surface area it presents to the shock wave relative to floor thickness. Progressive collapse, i.e., the failure of critical building supporting structures producing a cascading effect and ultimately total building destruction as it falls in on itself, can occur within seconds of the explosion.

When the overpressure shock wave radiates away from the explosive source, its magnitude decreases with distance and time and actually becomes negative as shown in Figure 5.6. This negative pressure causes suction forces on the building and the generation of a high velocity wind. The net result of this wind is to propel particles and debris creating an extreme risk of injury. Finally, nearby buildings are also subject to violent ground motion caused by the explosion, similar to short-duration earthquakes.

It is well known that blast effects are much different in confined spaces than in open areas. Confined explosions are accompanied by the shock wave reflections off the confining surfaces amplifying the explosive effect. In general, the result is more devastating for a confined scenario versus an open one, all other parameters being equal. This has significant implications for urban areas.

Explosive effects are actually quite complicated and developing a more precise picture requires computer modeling. Commercial programs using

Table 5.2 Explosive Delivery Methods and Corresponding Payloads

Explosive Delivery Method	Approximate Capacity (lb TNT)
Pipe bomb	5
Suitcase bomb	50
Automobile	500–1000
Van	4000
Truck	10,000–30,000
Semi-trailer	40,000

numerical techniques exist and only very rough calculations are presented here. However, these approximate results can still be useful for estimates of the vulnerability component of risk.

Since the two scenario-dependent parameters associated with explosive damage are the distance between the explosive source and the target plus the explosive payload, estimates on vulnerability require details on common methods of delivery combined with the potential damage inflicted by each method. Table 5.2 provides data on explosive payloads as a function of specific delivery methods. This information will help bound the security problem by developing physically realistic threat scenarios.

Although the effects of nearby structures are ignored greatly simplifying the analysis, Figure 5.8 is useful for estimating vulnerability to explosive threats since it shows explosive damage in terms of both distance and payload.[7]

This graphic depicts what I choose to call "lines of constant damage" for combinations of explosive payload and the distance between the explosive source and the target. In other words, each curve shows the combination of pounds of TNT-equivalent explosive (the US reference for explosive payload is pounds of trinitrotoluene, TNT) and the distance between source and target that produces the same physical damage.

So if there was concern about the vulnerability to the total destruction of a facility from a vehicle-borne explosion for a particular scenario, the bottom curve plots the combinations of distance and explosive payload that produce this effect. Clearly the closer the bomb is to the target the less payload is required to produce this effect. However, the salient question with respect to risk mitigation is how the explosive payload *scales* with distance for this destructive scenario.

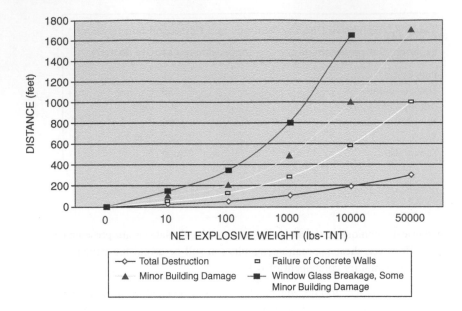

Notice that all four curves become increasingly "curvy" or nonlinear for increasing combinations of distance and payload. Increases in the distance and payload for a given damage scenario do not change proportionately for larger values of each parameter. For example, 100 pounds of TNT-equivalent payload detonated at a distance of almost 400 feet will produce the same damage (i.e., glass breakage) as 1000 pounds at 800 feet. So doubling of the distance from source to target forces an adversary to increase the explosive payload by a factor of 10 to cause an equivalent level of building damage. This illustrates the nonlinearly accruing benefits of increasing this distance, especially for large payloads. This is the type of analysis critical to assessing vulnerability, and it also demonstrates the value of graphics that provide visualizations of ballpark estimates.

Let's concoct a potentially familiar corporate scenario where Table 5.2 and Figure 5.8 might be useful in developing arguments on risk mitigation with respect to explosive threats. Suppose you are the security director of a high-profile, US-based international corporation and you have been asked to make a decision on the location of parking for your posh new company headquarters. This is the area that cars, taxis, limousines, etc., will use to pick up and drop off company executives and VIPs.

The senior managers of the firm, i.e., the people who pay your salary, are arguing vigorously to have the vehicle waiting area as close to the facility as possible. As the security director you have serious concerns about the wisdom of such a move. However, you must make a compelling argument to win the hearts and minds of those who are used to getting their way and who may view security as an annoyance if it interferes with their perquisites.

If we return to Figure 5.8 and examine the line of constant damage corresponding to "total destruction," we notice an interesting fact that would hopefully get management's attention. By increasing the distance between the explosive source and the facility by only 25 feet (i.e., doubling the distance to 50 feet) it would force a vehicle bomber to roughly increase the explosive payload by a factor of 10 to achieve the same destructive effect.

Putting things in stark but eminently practical terms, increasing the standoff in this way has the effect of eliminating a backpack-borne attack scenario (i.e., 10 pounds TNT equivalent explosive payload) and forcing an adversary to contemplate a vehicle-borne mode of attack (i.e., 100 pounds of TNT-equivalent explosive payload) to destroy the building. Adding 50 feet of separation or tripling the distance between a bomb and the target would force a would-be attacker to increase the explosive payload from 100 to 1000 pounds of TNT-equivalent explosives to destroy the facility.

We see that the added distance requirement imposed by moving the parking facility might take away the suitcase and possibly even the automobile-borne scenarios for explosive payload delivery if total building destruction is the ultimate objective. This is definitely a situation where a little distance goes a long way. It is also illustrative of how profoundly risk scales with physical parameters in this scenario. Often physical separation is the critical factor in determining the vulnerability component of risk.

Figure 5.9 represents a reformulation of the curves in Figure 5.8. In this version the vulnerability to building damage as a function of distance to a vehicle carrying 1000 pounds of TNT-equivalent explosive is presented. We readily observe that a separation distance of 800 feet is required to restrict damage to window breakage.

Unfortunately adversaries are often able to dictate the magnitude of relevant parameters in planning and executing an attack. However, this does not mean the defense is powerless in making reasonable estimates of vulnerability. In theory there are an infinite number of attack scenarios, but careful consideration of site-specific constraints, an adversary's objective(s), and an analysis of relevant physical parameters can yield valuable insights into the effectiveness of mitigation options. We will see a novel method of analyzing explosive vulnerability in the next section.

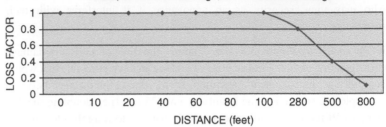

EXPLOSIVE DAMAGE MODEL FOR 1000 LBS TNT
(Car Bomb)
Loss Factors: 1 = total destruction, 0.8 = failure of concrete
walls, 0.4 = minor damage, 0.1 = window breakage

■ **FIGURE 5.9** Damage from 1000 pounds TNT-equivalent explosive.

5.3.2 **Confidence limits and explosive vulnerability**

A different approach to the problem of uncertainty in explosive attack scenarios has been utilized to better understand the vulnerability component of risk.[8] In this approach, a statistical formulation of scenario-driven explosive parameters has been created and is combined with a simple model for window explosive response. This enables the generation of confidence limits for window behavior under explosive loading with respect to the spectrum of explosive scenarios.

Specifically, this entails accepting the inherent randomness of both the explosive payload and distance from the explosive source in assessing the likelihood of window damage from vehicle-borne explosive threats. Debris from shattered glass is the largest contributor to injury and death in explosive incidents.

It is possible to exploit that randomness by modeling these parameters as Gaussian distributions with fixed limits determined by the particular scenario under consideration. This yields probability distributions for payload, $F_m(m)$, and distance, $F_r(r)$ as shown in Figures 5.10 and 5.11.

It is then possible to establish a joint probability distribution for payload and distance, $F_{mr}(m,r) = F_m(m) F_r(r)$ for the numerical values specified in $F_m(m)$ and $F_r(r)$ as shown in Figure 5.12. The joint probability distribution function displays a single maximum at (m_o, r_o) with a spread in the TNT-equivalent mass m determined by the dispersion in payload values, δm, and a spread in the stand-off distance, r, as determined by the dispersion in distance values, δr.

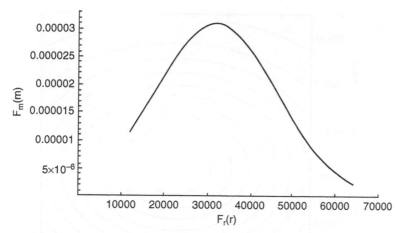

■ **FIGURE 5.10** TNT equivalent mass (pounds).

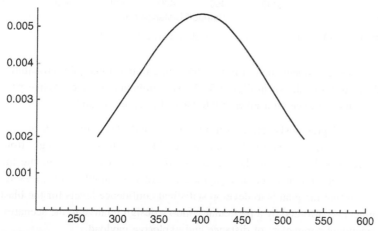

■ **FIGURE 5.11** Standoff distance (feet).

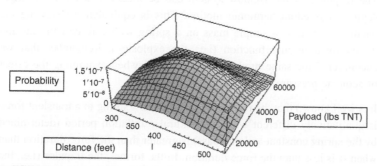

■ **FIGURE 5.12** Joint probability distribution for explosive payload and distance.

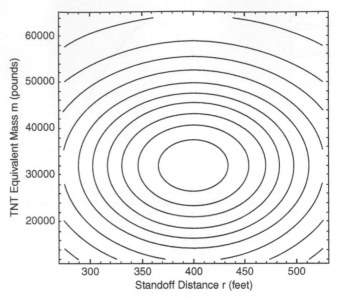

■ **FIGURE 5.13** Contour plot of the joint probability distribution for payload and distance.

A two-dimensional contour plot of this three-dimensional joint probability distribution is shown in Figure 5.13. All combinations of explosive payload and distance on a given circle have the same probability.

As noted previously, blast overpressure, p, and blast impulse, i, are the important quantities in determining building structural damage from explosives. We can use the scaling relations for each to reformulate the expressions for distance and payload in terms of impulse and pressure. Recall that the goal is to develop statistical confidence levels for the blast protection afforded by a particular window design relative to the scenario-dependent parameters of distance and explosive payload.

The response of the window system can be modeled in terms of single degree of freedom harmonic motion. This is equivalent to viewing the window system as a simple mass on a spring with a force constant and driven by a forcing function (i.e., the explosion). Remember that we encountered the same model in section 5.2.4 when examining the effect of acoustic pressure on walls.

It is well known that the response of a mass on a spring to a transient force is qualitatively different when the natural oscillation period (determined by the spring constant and the mass) is greater than the force duration than when it is less than the force duration. In the former, the impulse (i.e., the integral of the force over time) determines the initial velocity of response

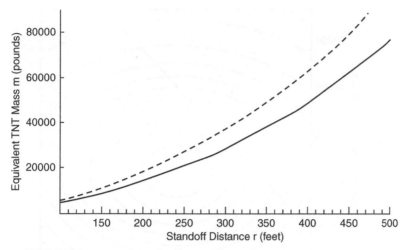

■ **FIGURE 5.14** Design curves for window blast protection.

of the mass, while in the latter the displacement of the mass is practically proportional to the force throughout its application. Different elements in a building have different natural oscillation frequencies, and the response of any particular element in its surroundings responds accordingly either to the blast overpressure exerted on it or to the blast impulse.

A unique design curve can be generated for specific values of pressure, impulse, and system natural oscillation frequency that can be transformed into design curves that are functions of payload and distance as shown in Figure 5.14. The solid curve has been chosen to pass through design specifications corresponding to pressure and impulse of 8 psi and 170 psi-ms, respectively, where the natural period of oscillation for the particular window under evaluation is 0.11 sec (56 sec^{-1} angular frequency). The dashed curve has been chosen to pass through design specifications corresponding to pressure and impulse of 16 psi and 185 psi-ms, respectively. Again 56 sec^{-1} is the window's natural oscillation period. The window will withstand combinations of distance and payload that lie below these values of the design curve.

Superimposing these design curves on to the contour plot of the joint probability distribution of explosive scenarios (Figure 5.13) and integrating the probability distribution F_{mr} (m,r) with respect to r and m reveals the probability of maintaining window integrity under explosive loading for the joint distribution of payload and distance. This is shown in Figure 5.15 and corresponds to an 80% confidence for the window designed to meet

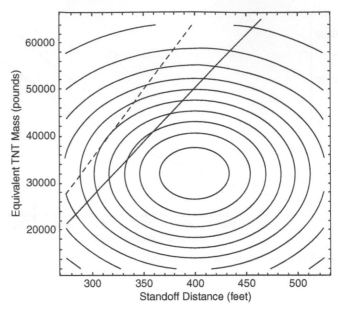

■ **FIGURE 5.15** Confidence limits of explosive blast protection for two window designs.

an 8 psi/170 psi-ms specification and 91% confidence for the window designed to the more robust 16 psi/185 psi-ms specification.

Windows represent a significant vulnerability to explosive incidents. So-called curtain wall structures are of particular concern and these are popular in modern commercial buildings. These structures consist of a system of metal framing with large glass windows as inserts. Effective curtain wall designs focused on addressing this threat have been developed for commercial facilities. Figures 5.16 and 5.17 illustrate the results of tests for one design that met or exceeded a US government facilities standard for a particular design basis threat. This standard specifies that a maximum of 10% of the exterior windows can suffer catastrophic failure (i.e., glass debris thrown in excess of 3.3 feet into the room) under explosive loading.

5.4 A THEORY OF VULNERABILITY TO COMPUTER NETWORK INFECTIONS

Computer viruses that propagate via the Internet continue to be an issue for most organizations. The computer security literature is replete with information on firewalls, virus scanning tools, etc., and this book will not address that well-documented area of security. However, there is less in the way of *measurable* risk indicators that might presage conditions that are ripe for the growth of network infections.

■ **FIGURE 5.16** Pre-explosive test curtain wall.

■ **FIGURE 5.17** Post-explosive test curtain wall.

Given the nearly universal dependence on communication via computer networks in office environments and the interconnectedness of machines on a worldwide basis, the ability to explicitly measure vulnerability to viruses has been a goal of computer security experts. Since the 1960s, there has been increasing attention to problems in network structure and organization. In particular, interest has been generated within the mathematics and physics community regarding theoretical work on network problems.

Some of that work has direct applicability to computer networks because of the use of computer networks as communication tools and features of interconnectivity. Specifically, the pattern of connectedness of virtual networks that evolve through e-mail traffic is believed to have a profound effect on the spread of infections.[9]

The computers of e-mail senders and recipients can be thought of as nodes of a network and form what is referred to as a social network. Other examples of social networks include connections between actors as well as the distribution of scientific references in journals. Social networks can be characterized as a small number of network members that have a large number of links to fellow members of the network, and a large number of members that have a small number of links to fellow members of the network.

In other words, a large number of people/nodes in the network communicate with a few people/nodes in the network whereas only a limited number of people/nodes communicate with many people/nodes in the network. One other important property relating to computer networks in particular is that they are constantly growing. Individuals in the virtual network formed by e-mail communications are connecting to additional nodes all the time, thereby continuously expanding the number of nodes in the network.

Figure 5.18 shows the probability of linking for nodes in two made-up e-mail networks. These are known as "scale-free" probability distributions and the equation that describes the linking is $P(k) = k^{-\gamma}$. The parameter k is the number of links per node (a link is defined as a connection between two nodes), and n(k) represents the number of nodes that possess a unique number of links. Recall from the earlier discussion on nonlinear functions that the minus sign in the exponent means the function is decreasing, and the rate of decrease is greater for increasing values of γ. So the magnitude of the exponent, γ, determines the steepness of the curve.

Observe that $P(k_1)$ is less steep (i.e., a smaller γ) than $P(k_2)$. Therefore $P(k_1)$ has more nodes with higher numbers of links. The higher number of links represents a risk factor with respect to the spread of network infections. What does this say about the likelihood of infection in the event of a computer virus? If I am a person who has a habit of sending e-mails to lots of friends and family members then clearly all those individuals are at risk if I get infected. Analogous with the world of infectious diseases, as the first infected individual, I would be referred to as the index case. However, if I generally send messages to only a few people then the probability of infecting other associates/network users is relatively low.

Therefore, the risk of infection by e-mail-borne viruses should be greater for the network node distribution $P(k_1)$ than the steeper $P(k_2)$ and the value of γ is an indicator of the risk of infection in the event of a computer virus. Lower values of exponents in scale-free distributions characterizing e-mail networks are at greater risk of viruses than distributions characterized by scale-free distributions with larger exponents.

It is the relatively small fraction of network nodes that communicate with the most number of other nodes that significantly affects the risk of infection in case of viruses or worms exploiting e-mail attachments. With respect to the graph in Figure 5.18, these nodes are at the far right of the horizontal axis (i.e., a high k value).

But network security types do not just sit around after an infection occurs. They typically jump into action and start repairing infected nodes. It turns out that the rate at which infected nodes are "cured" as well as the connectedness of such nodes affects the probability of the virus spreading. Specifically, the infection rate υ and remediation rate δ are key parameters, where the former represents how quickly an uninfected node becomes infected if it is connected to an infected node and the latter is the rate that infected nodes are restored to "health" following infection. The ratio of υ/δ is denoted by λ.

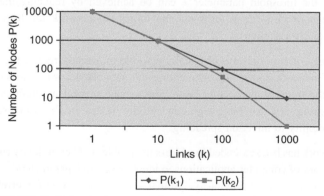

■ **FIGURE 5.18** Distribution of node connectivity for two (made-up) networks.

Incorporating these parameters leads to the identification of a threshold for computer virus propagation.[10] This means that a measurable parameter has been suggested that signifies a threshold condition for the spreading of computer viruses. In other words, if the network conditions are below that threshold then the infection does not grow. If network conditions exceed that threshold then the virus persists. This threshold condition has been identified as $D_2/D_1 < (1/\lambda)$ = the infection remediation rate divided by the growth rate where D_1 and D_2 are quantities derived from the distribution $P(k)$.

I noted previously that the steepness of the line characterizing the distribution of node connectivity is a risk indicator for the susceptibility of scale-free networks to viral infections. It turns out that the ratio of D_2/D_1 is proportional to the size of the network k_{max}, and the nodes with highest connectivity contribute most to infection growth. Therefore identifying, quarantining, and fixing those infected computers on a network that send e-mails to the largest group of machines on a network represents an expeditious security strategy in the event of an infection. Assuming individuals with lots of friends and colleagues do the most e-mailing, one method of enhanced vigilance might be to preferentially monitor those network users with the greatest number of entries in their list of e-mail contacts.

The results suggest that a scaling relation for risk exists with respect to the threat of computer virus infection. As noted previously, the smaller the value of the exponent γ in the distribution of linked nodes, the greater the number of highly connected nodes in a network and the greater the risk to infection by an e-mail-borne virus. Typical e-mail networks have been shown to have values of γ between 2 and 4, and in one case a measured value of 1.81 has been recorded.[11] According to the theory, it is easier to stay below the threshold for infection growth if γ is large. Staying below the threshold parameter λ can be achieved by reducing the size of the network since it turns out the larger the network the lower the threshold to infection.

Of course it is not always physically possible to actually reduce the number of network nodes. But segmenting the network according to subnets with the minimal number of nodes per subnet as well as preferentially monitoring and patching high-connectivity nodes might decrease the vulnerability to infection spreading.

This work has been extended to address the problem of "complexity risk" in a network of computer applications to identify key risk parameters[12] and is based on work cited previously.[8] In that study a collection of interrelated/interdependent computer applications is assumed to have a scale-free distribution. A rate equation for the probability that an application related to k other applications produces defective output is established using a so-called "mean field" approach. This establishes a network average for the probability that an application receives defective output from other applications.

Steady-state and time-dependent solutions are developed such that the probability of an application producing defective outputs is parameterized in terms of relevant variables. Conditions are then identified for receiving defective input from other applications. In addition, a threshold for which the network of applications will be susceptible to appreciable input errors is identified as well as the dependence of defective input on the number of interrelated applications in the network.

5.5 BIOLOGICAL, CHEMICAL, AND RADIOLOGICAL WEAPONS

5.5.1 Introduction

In the last few years the types of physical threats on the radar screens of security professionals have broadened. In the not too distant past, vandalism, theft, robberies, and general harassment constituted the domain of corporate security professionals. Accordingly, locks, alarms, and guards represented the staple of required antidotes. Although it is unclear just how much the likelihood component of risk associated with nontraditional threats such as biological, chemical, and radiological weapons has increased, their imprint on the public psyche has clearly grown.

There have been documented incidents of the use of these weapons, especially in war, dating back to World War I. Since then a number of countries have created and maintained active research programs in this type of weaponry. Although there is little history of terrorist organizations utilizing these methods, this might not be due to a lack of interest on their part. Materials required for such weapons might be stolen from unprotected civilian institutions and the possibility that these materials might find their way into the hands of less reputable characters is concerning to say the least.

In general these types of attacks would be difficult to spatially confine, so even if a particular facility were targeted there is a risk of collateral effects. Therefore, the prudent security manager must gauge the likelihood and vulnerability to possible threat scenarios and plan accordingly. It is one thing to make an educated estimate of risk and then develop a cost-effective strategy, which might include doing nothing. It is quite another story to ignore the problem entirely or merely guess at an appropriate mitigation strategy leaving the outcome to chance.

5.5.2 **Vulnerability to radiological dispersion devices**

Let's begin with what some experts consider to be one of the easiest non-traditional attacks to implement, the radiological dispersion device (RDD) or "dirty bomb." In this scenario, a conventional explosive is detonated along with radiological material. This would cause the radioactive material to break up into smaller radioactive chunks that would spread over an area with dimensions determined by the explosive payload and local environmental conditions (e.g., wind).

An RDD is not the same thing as the nuclear bomb used on the cities of Hiroshima and Nagasaki during World War II. Those devices used nuclear fission to release tremendous amounts of explosive energy equivalent to thousands of tons of TNT. The vulnerability to the explosion of an RDD is determined in the same way as conventional explosives. The twist here is that an RDD combines a conventional explosive with a radioactive isotope giving the terrorist a twofer in terms of nasty effects.

It is a fact of life that humans are constantly exposed to a background level of radiation and unfortunately the effects are cumulative.* Some of this radiation derives from primordial sources and some rains down on us from extraterrestrial sources such as the sun. Moreover, there are commercial products and devices that utilize radioactive sources. These include smoke detectors, certain types of gauges, sources of medical irradiation, etc.

Radioisotopes produce particles and/or electromagnetic radiation as part of their natural process of nuclear decay. These materials typically emit radiation in the form of gamma rays, beta particles, and/or neutrons, affecting individuals in relatively close physical proximity. Unfortunately, radioactive emissions are invisible but can be very deleterious if inhaled or given enough exposure via a combination of radioactivity, physical proximity, and time.

Isotopes of an element are differentiated by varying numbers of neutrons in the atomic nucleus. These are labeled by a superscript next to the element symbol (e.g., ^{60}Co), where the superscript denotes the sum of the number of protons and neutrons in the atomic nucleus. Radioisotopes are inherently unstable and undergo a continuous process of nuclear decay. Although this section is not intended as a primer on radioactivity, there are scientific concepts that are necessary in estimating the vulnerability component of risk associated with the threat of RDDs.

---

*An excellent exposition on background radiation and nuclear energy in general can be found in "Megawatts and Megatons", Garwin and Charpak, University of Chicago Press, 2002.

The first concept to understand is the notion of nuclear activity. The traditional unit of measurement of activity is the curie (Ci). A curie is defined as 3.7×10^{10} nuclear disintegrations per second. The amount of radioactive material that is present is specified by the so-called specific activity or activity per mass.* A thousand curies of substance A and a thousand curies of substance B are equivalent from a radioactive perspective, although they will not necessarily have the same mass unless they have the same specific activity. The modern unit of nuclear activity is the Becquerel (Bq) and is defined as one nuclear disintegration per second. Therefore, $1 \text{ Ci} = 3.7 \times 10^{10} \text{ Bq}$.

Gamma radiation, beta particles, and neutrons all have deleterious biological effects on human tissue if thresholds of exposure are exceeded. A unit of measurement exists that normalizes these effects so that a comparison of dosages can be made for various forms of radiation. This unit is called the Roentgen Equivalent Man or rem. The modern dose equivalent unit is known as the Sievert (Sv), where 100 rem = 1 Sv.

The US Health Physics Society has specified that clinical effects are not observable in humans for dose equivalent levels below 35 rem. Serious short-term effects occur at 70 rem, and death occurs at roughly 400 rem. In case you are concerned about having been recently zapped at your last trip to the dentist, a person receives a dose equivalent of about three thousandths of a rem from a dental X-ray and about 0.4 rem/year is contributed from natural background radiation. As noted above, each person on earth is continuously exposed to low levels of nuclear radiation and there is absolutely nothing we can do about it. In fact, we are continually irradiating ourselves due to the minute presence of radioactive material in our own bodies.

There are a large number of radioisotopes and many could potentially be deployed in an RDD. However, the availability, toxicity, and persistence of the material are factors of key importance to anyone considering committing havoc in this way. By persistence I mean that the substance remains radioactive for a sufficient period of time to cause ongoing angst among the target population. A measure of persistence is the radioactive "half-life." This is defined as the time required for the population of continuously decaying radioactive nuclei to be reduced by half.

Another way of understanding the risk of exposure to radioactivity is by considering a metric that specifies a rate of 0.6 additional cancers per 10,000

---

*Radioactive decay is a random process and obeys Poisson statistics (see Chapter 4). Therefore, the exact arrival time for a specific decay event can not be predicted but the probability of a number of counts occurring in a given time interval can be calculated.

people-mSv. This says that if 10,000 people experience a dose equivalent of 1 mSv, the group could expect to see 0.6 additional cases of cancer.

Three radioisotopes that are relatively easy to obtain and might have attractive features for a socially distorted adversary are Cobalt-60 (^{60}Co), Iridium-192 (^{192}Ir), and Cesium-137 (^{137}Cs). Each produces gamma radiation with different energies. The biological effect from radiation as a function of activity can be measured in units of rem/curie-hour at one meter or RHM. This is a useful parameter in assessing risk associated with the threat of an RDD since it takes into account the deleterious biological effects due to the total time of exposure, distance from the source, and source quantity.

The US Health Physics Society provided a metric that is useful in assessing vulnerability to RDDs. They specify that 1000 Ci (about 11.4 g) of ^{137}Cs spread evenly over a circular area of radius 100 m ($\sim$7.8 acres) could produce observable clinical effects after 100 hours of unshielded exposure. Based on the Health Physics Society metric, I assume 35 rem is the dose equivalent threshold for significant health concerns. Values of around 50 rem (0.5 Sv) are generally cited as the minimum threshold for the onset of symptoms of radiation poisoning.

The 35 rem US Health Physics Society figure tells us that roughly four days of exposure to an exploded 1000 Ci source of ^{137}Cs would be required before clinical effects from the associated radiation would be seen. It is clear that this would not be a particularly happy day for anyone who was so exposed, since any unnecessary exposure to radioactive material is unhealthy. But the acute vulnerability to the threat of an RDD seems to be principally from the initial explosion rather than the radioactive emissions from the isotope. Extended exposure to that isotope would be required to cause significant health problems. However, the health effects can be significant if the radioactive material makes contact with the skin or is ingested.

It is useful to evaluate the vulnerability to the 35 rem dose equivalent threshold via a small chunk of unshielded radioisotope as a function of distance from the source. In all likelihood this would not be an accurate representation of an RDD scenario, since in real life the radioactive source would be dispersed by the explosion. Nonetheless it is illustrative in examining the vulnerability component of risk. The RHM values for ^{60}Co, ^{192}Ir, and ^{137}Cs are Cobalt^{60}, are 1.35, 0.59, and 0.38, respectively.

Assume there was concern about a 1000 Ci radiotherapy source of ^{60}Co (about 0.9 g) that was reported missing from a local hospital. The vulnerability to radiation is a function of the substance activity (i.e., the number of curies), the type of emitted radiation, the effect of intervening material, and the distance between the substance and a vulnerable individual.

A 1000 Ci source of ^{60}Co would cause a 1350 rem dose equivalent in 1 hour (1.35 rem/hour-curie $\times$ 1000 curies) at a distance of 1 meter. That means that in about 1.6 minutes a human would be exposed to the clinically safe limit for an entire lifetime. ^{60}Co is particularly nasty stuff. It should be noted for a 2000 Ci source of ^{60}Co the time required for the same dose equivalent would be cut in half. To estimate vulnerability, the familiar image of the expanding sphere as depicted in Figure 2.8 is again invoked to understand how the intensity and associated dose equivalent change with distance. Since radioactivity is emitted equally in all directions from a small chunk of material, and we are ignoring the effects of intervening material for now, we can expect a $1/r^2$ dependency of intensity with distance where r represents the distance between the source and an intervening object. This should sound familiar.

The flux of radioactivity hitting a human body from a chunk of ^{60}Co is the number of gamma rays hitting the body per unit area per unit time. Therefore the number of gamma rays hitting a unit area of this sphere per unit time decreases as one gets further from the source. As the radioactive flux decreases so should the absorption by a target of opportunity. However, quantifying the effects of radioactive absorption in humans is more complicated than just calculating the radioactive flux, since it depends on the organ irradiated and other biological factors.

But as I have gently hinted, this is not the whole story in understanding the vulnerability to a radioactive threat. We are surrounded by an envelope of air and this envelope is also an absorber of radiation. The gamma rays emitted from the chunk of ^{60}Co interact with the atomic elements of the nitrogen (78%), oxygen (21%), etc., that comprise air. Although the interaction of radiation with materials is complicated, the governing phenomena are well understood by physicists and are functions of the type and energy of the ionizing radiation as well as the atomic number of the material.

The attenuation of electromagnetic energy such as gamma rays can be calculated using a parameter that accounts for all the attenuating processes. It specifies the decrease in intensity as the radiation plows its way through a given material. This parameter is known as the mass attenuation coefficient. With respect to gamma radiation traveling through air at the 1 MeV energies of the two decay products of ^{60}Co, the mass attenuation coefficient is approximately equal to 0.1 cm^2/gm.

Figure 5.19 indicates the mass attenuation coefficient for gamma/X-rays propagating in air over a range of energies (solid line).

The intensity of ionizing energy as it passes through material obeys a simple differential equation and an equation of this type was introduced in Chapter 2. The intensity of a beam of radiation as a function of distance, r, as it traverses

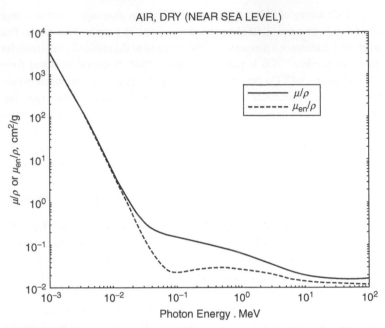

AIR, DRY (NEAR SEA LEVEL)

■ **FIGURE 5.19** Mass attenuation coefficient for X-Rays and gamma rays in air. *From http://physics.nist. gov/PhysRefData/XrayMassCoef/ComTab/air.html.**

through the material decreases at a rate determined by the value of the mass attenuation coefficient, μ/ρ, where ρ is the density of the material and μ is the linear attenuation coefficient.

Specifically, the mass attenuation coefficient** determines the total loss of intensity of a beam of electromagnetic energy traversing matter due to the effects of both absorption and scattering. With respect to gamma and X-rays, absorption and scattering result from a combination of the photoelectric effect, pair production, and/or Compton scattering, and the relative contribution of each process depends on the beam energy and the material. Pair production is not relevant for gamma energies below about 1 MeV.

The first order differential equation that describes this process of attenuation is

$$dI/dr = -(\mu/\rho)I \tag{5.9}$$

---

*The curve labeled μen in Figure 5.19 represents the energy-absorption coefficient which is the fraction of incident energy locally absorbed per centimeter.

**The mass attenuation coefficient should not be confused with the mass energy-absorption coefficient. The former gives the photons interacting per unit mass of medium rather than the energy absorbed. The mass attenuation coefficient is used in the calculation of the number of photons that reach a point while the mass energy-absorption coefficient is used to calculate the absorbed dose of radioactivity once the photons arrive.

We also know from Chapter 2 that the solution to that equation is a decreasing exponential and is given by

$$I = I_0 e^{-(\mu/\rho)\rho r} \tag{5.10}$$

I_0 is the initial intensity, e is the exponential ($\sim$2.72), μ/ρ is the mass attenuation coefficient, and r is the distance the gamma ray energy travels through the material. Since μ/ρ incorporates density in the denominator and has units of cm^2/gm, one must multiply by the density of the material (in this case air), ρ, in the exponent. Details of this calculation are provided in Chapter 6 in the box titled, Security Metric: Lead Shielding of Gamma and X-ray Energy.

Using this expression combined with the $1/r^2$ decrease in intensity due to isotropic radiation from a point source, the attenuation of gamma energy can be calculated to arrive at a metric of vulnerability to health effects from the threat of an RDD. Using the values for the specific activities noted previously, Figure 5.20 shows the time required for a 35 rem dose equivalent of radiation as a function of distance through air from 1000 Ci of ^{60}Co, ^{192}Ir, and ^{137}Cs.

We see from the curves that the required time for 35 rem dose equivalent dramatically increases for distances greater than 50 m for all three isotopes.

Since the radioactive nuclei are continuously decaying, you might be tempted to believe you could just wait out the decay process and hang around as the chunk decays to a harmless mass. Unfortunately this time Mother Nature is working on the side of the adversary depending on the isotope. In fact, a thoughtful adversary might even choose a particular isotope in part for its radioactive persistence.

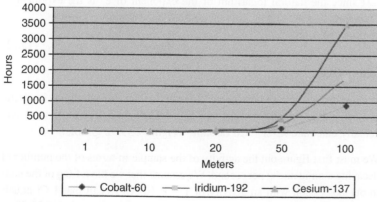

■ **FIGURE 5.20** Time required for 35 rem dose equivalent as a function of distance from a 1000 Ci radioactive source.

I previously mentioned the term half-life, which characterizes this persistence. It is defined as the time for the population of radioactive nuclei to be reduced by one half. This important concept is worth exploring in more detail.

Once again we encounter a physical process where the rate of change of intensity with respect to time is proportional to the negative of the intensity itself (it is negative because the intensity is decreasing with time), and where the proportionality constant is specified in terms of a process-related rate constant. Let's denote this rate constant using the Greek letter lambda, λ.

We know from similar physical processes that the solution to the equation and resultant radioactive intensity includes the exponential, e, and that the intensity decays with time according to

$$I(t) = I_0 e^{-\lambda t} \tag{5.11}$$

I, I_0, and e are the intensity, initial intensity, and the exponential, respectively. λ is the decay constant for this material, and t is time. Let's solve for t when the ratio of the intensity relative to the initial intensity I_0 is ½. In other words, we seek the time t, such that the intensity of the radioactivity from the radioisotope has been halved in value, i.e., the radioactive half-life. Fortunately, we already know how to calculate this.

In Chapter 2 we also learned that the natural logarithm's function "undoes" the exponential function, e^x. In other words, $\ln(e^x) = x$. So, we take the natural logarithm of each side of equation (5.11). On the left side we perform the natural logarithm of I/I_0 or $\ln$ (½) since we are interested in when the ratio of the intensity to initial intensity (I/I_0) is ½. This calculation yields -0.69.

Performing the same operation on the right side of the equation we obtain $-\lambda t$ since the natural logarithm of the exponent of e^x is the exponent x. Therefore,

$$\ln(1/2) = -0.69 = -\lambda t \tag{5.12}$$

so, $t = 0.69/\lambda$.

If we stick with the radioisotope ^{60}Co we could merely look up its decay rate constant, λ, to arrive at the half-life, t. Let's do a little more work and actually calculate λ for ^{60}Co (atomic weight = 59.92 g/mole).

We must first figure out the activity of the sample in terms of the number of decaying atomic nuclei per second. We assume that we have 1.0 g of the stuff in our possession with an activity of 1.1×10^3 Ci (recall that 1 Ci equals 3.7×10^{10} atomic disintegrations/second). This implies 1.1×10^3 Ci $\times$ 3.7×10^{10} decaying atoms/Ci-sec = 4.1×10^{13} decaying atoms/second.

We must now calculate the number of atoms in the 1.0 g sample as follows:

$(1.0$ g$)/(59.92$ g/mole$) \times (6.0 \times 10^{23}$ atoms/mole$) = 1.01 \times 10^{22}$ atoms.

So the decay constant is given by $(4.1 \times 10^{13}$ atoms/sec$)/1.01 \times 10^{22}$ atoms $= 4.1 \times 10^{-9}$ sec$^1 = \lambda$.

Plugging this into the expression for the half-life, t, of ^{60}Co we get 0.17×10^9 sec or 5.39 years.

If we assume the half-life of ^{60}Co is roughly 5.4 years, this means in 5.4 years half of its radioactive nuclei will have decayed by emitting gamma rays (the two distinct gamma rays of have 1.17 and 1.37 MeV energies).

Suppose a person is remarkably self-destructive or equally crazy and intentionally picks up a chunk of ^{60}Co. By some miracle he survives this first close encounter as well as a second one after 5.4 years. "Better lucky than smart" is the operational paradigm here. Although our radiologically challenged individual would probably not survive either experience, a measurement of the radioactive intensity of this material would be found to be half of what it was when he first picked it up 5.4 years ago.

And suppose this same individual did not learn from his mistakes and he picked up the same chunk again after another 5.4 years. This time the intensity would be half of what it was in the previous 5.4 years and so on. Every 5.4 years the radioactive intensity would decrease by half its value. This rate of decay of the intensity of radioactive substances is characteristic of exponentially decreasing processes. It is unfortunate that humans do not decay in a similarly graceful fashion.

Therefore, it seems we would have to wait a long time for the intensity to diminish sufficiently to be safe from radioactive exposure. What if an adversary chose a different isotope? The half-lives of ^{137}Cs and ^{192}Ir are 30.1 years and 74 days, respectively. From the adversary's point of view, it is important to use a radioisotope of sufficient longevity since the principal goal is to cause disruption for as long as possible. But as we have seen, close and sustained proximity to the chunks of dispersed radioactive material is required to cause significant biological damage. It is worth repeating that the explosive is the most dangerous feature of an RDD and I have shown why this is so.

So far the calculations of radioactive intensity have assumed point sources of radioactivity for the dispersed RDD material. How would things change if we modeled the decay using physically larger sources? Radiation from an RDD will probably be scattered quasi-evenly over some area. We might approximate the shape of that area as a disc. The intensity of

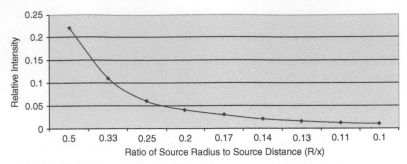

■ **FIGURE 5.21** Radiation intensity as a function of distance from an extended radioactive source.

radiation from a disc-shaped source of radius R at a point p located a distance x from the disc is given by[13]

$$I = (S/4) * \ln(1 + R^2/x^2), \tag{5.13}$$

S is the number of photons (i.e., gamma rays) emitted per area per unit of time. Figure 5.21 shows how the relative intensity varies as the ratio of the source radius to the distance to the source (i.e., R/x), neglecting the attenuation due to air.

When the distance to the disc, x, is much less than the radius of the disc, R, the intensity at a point, p, is given by[14]

$$I = (S/2) * \ln(R/x) \tag{5.14}$$

In summary, the features influencing the risk associated with radiological threats are the distance from the radioactive source, the energy, type of emitted radiation, the amount of radiological material, the exposure time to the radioactive source, and the attenuating effects of any intervening material.

There is ongoing concern for the effects of RDDs on facilities and specifically over the radioactive detritus in the streets that might find its way inside buildings. We will see that contaminated air can enter buildings through the façade and this is a problem with respect to the vulnerability of a building population. The building air intake vents represent an obvious vulnerability to an RDD since these are designed to suck in large volumes of external air.

Following the explosion of an RDD, the resulting radioactive cloud expands in radius, rises gradually to a maximum height, and drifts downwind. For radioactivity from the cloud to enter an open building vent, the vent must be enveloped by the cloud sometime during its expansion, rise, and drift. It would be useful to have an estimate of the vulnerability of the vents to envelopment by the radioactive plume as a function of height above the ground and distance from the explosive source. In one unpublished study[15] the key parameters affecting explosive material enveloping building vents were analyzed for a few representative payloads in an attempt to more precisely understand the vulnerability to RDDs. Only the van-sized delivery mode is discussed here.

Specifically, the effects of both the magnitude of an explosive payload and the distance between the explosive payload and a facility were examined in determining the likelihood that by-products of that explosion would reach elevated air intakes. In addition, the time dependence of this process was calculated to determine the time available to close building air intakes.

This study showed that the time required for a radioactive cloud to envelop-vents located 300 m above the ground ranged from seconds to a few minutes at most. In addition, a bimodal structure appeared for a van-sized bomb (~5500 pound TNT equivalent): the vents are enveloped for a range of small downwind distances x between explosive source and facility as well as a range for large x. These two regions of distance are shown in Figures 5.22 and 5.23 where the vertical axis represents radioactive cloud height (cm) and uA is radioactive cloud velocity (cm/sec).

These figures show that the radioactive cloud can envelop highly placed vents at very small downwind distances. The envelopment at small distances occurs before the cloud has risen so high that the vents lie below the bottom of the cloud. Also, for intermediate explosive payloads (i.e., explosive payloads that might be carried by a van-sized vehicle), the radioactive cloud can envelop the highly placed vents at rather large downwind distances.

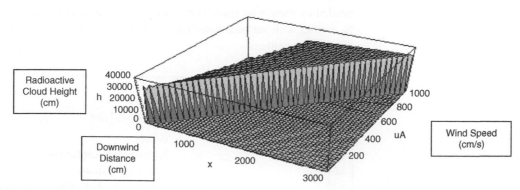

■ **FIGURE 5.22** Envelopment of elevated building vents at small downwind distances x.

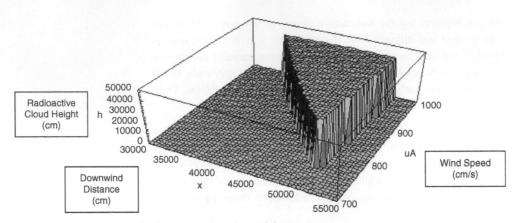

■ **FIGURE 5.23** Envelopment of elevated building vents at large downwind distances x.

The envelopment at large downwind distances occurs when the radius of the cloud has grown so large that the bottom of the cloud extends lower than the vents, despite the fact that the cloud's center of mass is very high above the ground.

5.5.3 **Vulnerability to biological threats**

Biological threats are a relatively new concern to the civilian security sector. These threats come in three basic types: bacteria, viruses and toxins. Table 5.3 shows the type and range of diameters of some general types of pathogens. This could be useful in understanding vulnerability to building contamination in conjunction with data on building HVAC filter performance.

The delivery mechanism envisioned for these threats is via an aerosol spray, where atomized particles are released into the atmosphere to be inhaled by those who come in contact with the toxic plume or picked up by contact with surfaces upon which the pathogen has settled following droplet evaporation.

Table 5.3 Diameter of Pathogens

Diameter (μ)	Pathogen Type
< 0.3	Viruses
0.3–1.0	Bacteria, dust, Legionella
3.0–10	Dust, molds, spores
>10	Light pollen, dust mites

It is also possible that for some pathogens the disease would be contracted from those who were infected by the original plume, although typically this will not occur until later in the incubation period for viruses. Symptoms may not manifest in individuals until days after the attack has been initiated complicating response options.

A reasonable assumption about a biological attack seems to be that it will be initiated at ground level. Another is that biological agents would be dispersed using some form of aerosolizing mechanism rather than conventional explosives. Biological material delivered by an explosive device might be possible, but this mode of delivery could be tricky to implement since the high temperatures of detonation might kill the pathogens.

Therefore, a significant building mitigation feature for the dispersal of biologically based threats is elevated air intakes given the assumption that an aerosolized attack is likely to be initiated at ground level and material dispersed would be heavier than air. Physical access to air intakes, especially those located at street level, represents a heightened vulnerability to this threat since these afford direct access to the air circulating within a building.

I will discuss potential mitigation strategies for biological threats in more detail in the next chapter, so I will temporarily postpone a more detailed discussion of filtering. However, it is obvious that the vulnerability to biological agents may be impacted by the type of defensive measures deployed within a building's heating, ventilation, and air conditioning (HVAC) system, since this is constantly pulling in large volumes of air from the outside (and potentially contaminated) environment.

In accordance with the hypothesized aerosol-based attack mechanism, it is reasonable to postulate that some form of nebulizer would be used to produce the aerosol. These devices generate droplets of liquid material and are familiar to those who water plants, clean the house, etc., although presumably terrorists would use more powerful versions than the plastic hand-held variety I have in my apartment.

For such an attack it might be important to know how long the droplets remain in a liquid state. In other words, we want to know the droplet evaporation time since the effectiveness of mitigation methods could depend on whether the toxin is in a liquid or solid state. In addition, a mitigation strategy might be preferentially oriented to address infection by either inhalation or surface contact to optimally suppress the rate of growth in the population.

Specifically, biological material that has been dispersed via liquid droplets sprayed into the air would presumably exist as a salt upon droplet evaporation. Droplets sprayed inside a building would be subject to the inertial forces exerted by the building air circulation system. Therefore, lower system air velocities would have a more limited effect in terms of risk from inhalation since droplets would be carried less distance; hence evaporation would take place over a smaller area.

A calculation has been performed to determine droplet evaporation time[16] and I quote directly from these results. A spherical droplet of water or other material with a vapor density, C_o, was assumed for this calculation. This is the density of the saturated vapor in g/cc at 20°C and a vapor pressure of 17 mmHg so that the water vapor density is equal to 1.8×10^{-5} g/cc. The vapor has been assumed to have a binary diffusion coefficient of D in air, and the liquid is assumed to have a density ρ, which in the case of water is 1 g/cc. A reasonable handbook value for D is 0.2 cm^2/sec.

The scaling relation between the droplet radius and the time before evaporation is as follows:

$$r^2 = 2DC_o(t_o - t)/\rho. \tag{5.15}$$

Therefore, the time to evaporation, t, scales as the square of droplet radius. Below about 0.1 μ droplet radius, where free-molecule flow takes over, the evaporation rate was shown to be simply proportional to the product of the vapor density and the free-molecule speed. In the region of larger droplet sizes, which is the region of concern for biological threats, the results are graphically depicted in Figure 5.24.

Time (ms)	1	3	10	30	100
Droplet radius (μ)	0.85	1.47	2.68	4.65	8.49

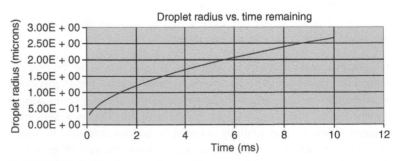

■ **FIGURE 5.24** Droplet radius versus time remaining as a droplet

A 17-μ diameter droplet would last 0.1 sec, and in an air flow of 500 cm/ sec, would be carried 50 cm downstream. A 3-μ diameter droplet would be carried for 3 ms or a distance of 15 cm in that same air flow. Therefore, the diameter of the droplets in question has implications for the probability of airborne versus surface-borne infection and the effectiveness of building filtration.

The vapor pressure of a very small droplet is enhanced because of the capillary pressure (i.e., the difference in pressure across the interface of two immiscible fluids...think surface tension). It is reduced by the electric tension of the surface charge. The tension cannot exceed the capillary pressure without destabilizing the droplet. Capillary pressure is important for droplets in the nanometer size range. Moreover, charge tension can be comparable with capillary pressure only for droplets far larger than those of concern for biological threats.

Although the air intakes represent a significant vulnerability to the threat of internal building contamination due to external sources of nasty substances, shutting the system down and closing the vents will not necessarily be an effective mitigation strategy. Ultimately the effectiveness depends on the building design, wind speed, external versus internal temperature, and the timescale of interest.

The effectiveness of filtering and therefore the vulnerability to airborne pathogens is also a function of the air tightness of the building skin. If a building is porous to outside air, the filtration will be somewhat negated since unfiltered air can enter the building by bypassing the air intakes. Buildings that are well-insulated from the outside and use high quality filtration in conjunction with positive pressurization relative to the external environment are better protected against an external biological attack. But recognize that positive pressurization requires clean air to be taken from somewhere. Positive pressurization with clean air is nontrivial to implement in a commercial facility if the external environment is contaminated.

Standard commercial buildings are never completely airtight. In fact many buildings are not well-insulated at all, especially older ones. Airtightness must be determined by performing a pressurization test on a floor-by-floor basis. One general specification for airtightness is the room air exchange per hour (ACH). This is the rate at which outside air replaces inside air within a room. In the absence of an active HVAC system, the inside and outside air concentrations ultimately reach an equilibrium state. Natural ventilation rates range from 0.25 room air exchanges per hour for tightly constructed buildings and 2.0 room air exchanges per hour for less tightly constructed facilities.[17]

An estimate of vulnerability to various pathogen types can be made by matching up filter effectiveness as a function of pathogen diameter. However, even if air filtration is present it is recommended that the HVAC system be shut off in the event of a known external air contamination incident given the tremendous flux of external air pulled into the facility by the HVAC system. Building occupants are often advised to "shelter-in-place." This is intended to provide a period of relative safety since buildings theoretically offer some degree of protection from an external attack with windows, vents, doors, etc., closed. Or do they? How much time do building occupants have before they too are vulnerable to external contaminants?

An estimate of vulnerability associated with this strategy can be calculated based on the airtightness specifications previously quoted and an assumption that the change in concentration of air with time obeys a simple and now familiar dynamic. Let's assume there is a concern for an external contamination incident and the HVAC has been shut off. Initially there is no contaminated air in a room that is contiguous with the outside.

We know that over time contaminated air will seep into the room from the external environment until the concentration of contaminated air inside the room equals the concentration outside of the building. We will assume that when half the room air consists of outside air it is no longer considered safe for occupancy. Of course this supposition depends on the type of attack and is intended only as an illustration.

Internal and external air is exchanged across small fissures in the building facade. The concentration of air as a function of time can be assumed to behave according to the familiar simple first order differential equation first introduced in Chapter 2; namely, the change in concentration of clean air in a room with time equals the decrease in the concentration of clean air times the exchange rate of air with the outside environment, R, and is written as follows

$$dC/dt = -RC \qquad (5.16)$$

Based on this process of air exchange and assuming two room exchanges per hour or a leaky building rate constant, 50% of the inside room air is contaminated by outside air in about 21 minutes (see the security metric calculation "Room Air Contamination Time").

We know that the solution to this equation is a decreasing exponential, so that for each successive 21-minute period, half of the remaining inside room air would be exchanged with the outside environment. If we used a figure of 0.25 air exchanges per hour, i.e., the rate constant corresponding to an airtight building, then the time for half the room air to become contaminated by outside air is 2.8 hours.

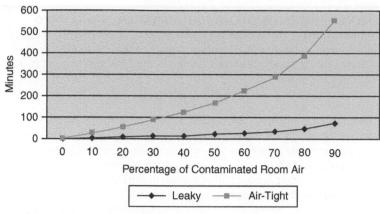

■ **FIGURE 5.25** Room contamination time.

Consider how this might affect a proposed shelter-in-place strategy if you were depending on the building to offer protection to its occupants. Figure 5.25 graphically illustrates room contamination times for leaky and airtight buildings based on Equation 5.16:

SECURITY METRIC: ROOM AIR CONTAMINATION TIME

The rate of exchange of outside air with inside air can be written simply as above

$$dC/dt = -RC \tag{5.17}$$

where C represents the concentration of non-contaminated (i.e., air that is inside the facility when the contamination event occurs) air in the room, R is the rate of air exchange with the outside, dC/dt is the change in concentration of inside air with respect to time, t. The quantity RC is negative because the concentration of clean or non-contaminated air (i.e., inside air) is decreasing with time. This is a very common form of equation for describing natural phenomena and similar forms of the same equation abound in nature (e.g., radioactive decay).

The solution to the equation $dC/dt = -RC$ is given by

$$C = C_0 e^{-Rt} \tag{5.18}$$

where C_0 is the initial concentration of (clean) inside air. We want to know the time it takes for half the initial inside air to be contaminated with outside air. You will no doubt notice that this is shockingly similar to the expression for radioactive half-life:

$$C/C_0 = \frac{1}{2} = C_0 e^{-Rt} \tag{5.19}$$

(Continued)

SECURITY METRIC: ROOM AIR CONTAMINATION TIME—CONT'D

Recall once more that we can "undo" the exponentiation process by taking the natural logarithm of the exponential function; hence we must do the same to the left-hand side of the equation. In this case, the exponent is −Rt. This yields the expression

$$\ln(1/2) = -0.69 = -Rt \tag{5.20}$$

Knowing R allows one to solve for the time t. If the rate constant R is assumed to be that of a leaky building or 2 room exchanges per hour, then t is approximately 21 minutes. The fraction of outside air required to produce harmful effects to humans depends on the contaminant, and if this is known, one could substitute that fraction instead of ½ to arrive at the contamination time of concern.

Using the expression for dC/dt allows us to assess the vulnerability component of risk associated with threats from external contamination. Again, I have assumed that the building HVAC system is shut off and I made an educated and worst case guess on the rate constant, R. This simple estimate could be quite important in developing a realistic shelter-in-place strategy in the event of external particulate, vapor release, RDD, or other contamination-type threat.

5.5.4 **Vulnerability to external contaminants; bypassing building filtration**

For a number of buildings the opportunity to shelter-in-place in response to external contaminants is problematic even with the HVAC operating unless positive pressure can be maintained relative to the outside. There is good reason to believe that many building façades exist in varying states of leakiness. This feature makes them susceptible to the so-called stack effect, which jeopardizes the effectiveness of any filters or sterilization methods installed upstream of the external HVAC air intakes.

The stack effect (also known as the chimney effect) results from the creation of internal vertical temperature gradients that produce pressure differences with the outside environment[18] If the density of air inside the building as a function of height above the ground is different than the pressure existing on the outside of the building at the corresponding elevation, air will flow in or out of the building depending on the relative pressure difference (i.e., greater inside the building than outside or vice versa).

The presence and distribution of holes in the building exterior as well as the unobstructed vertical flow of air within the building interior are major factors affecting the magnitude of the stack effect. This in turn affects the

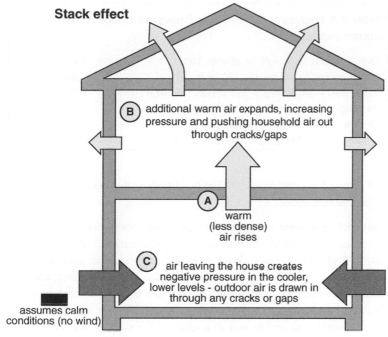

Stack effect

(B) additional warm air expands, increasing pressure and pushing household air out through cracks/gaps

(A) warm (less dense) air rises

(C) air leaving the house creates negative pressure in the cooler, lower levels - outdoor air is drawn in through any cracks or gaps

assumes calm conditions (no wind)

■ **FIGURE 5.26** The stack effect. *Courtesy of Superior Chimney Services Corporation.*

amount of unfiltered air entering the building even with the HVAC filtration/sterilization systems operating. A diagram illustrating the effect in a house is shown in Figure 5.26.

Figure 5.26 illustrates the case where the air inside the building is heated, which is the case in winter months in the Northern Hemisphere. Warm air inside the building rises to upper levels since the buoyant force of the volume of heated air exceeds the gravitational force exerted on the same volume of air. This causes an increase in air pressure at higher elevations and a simultaneously reduced pressure at lower elevations. Air at the upper levels is therefore "pushed" out of cracks in the exterior since the pressure there is greater than it is on the outside of the building at that height.

In the absence of other sources of pressure, the amount of air exiting the building at upper levels exactly balances the amount of air drawn in at lower levels. In summer months the effect is reversed, as denser, air conditioned air at the lower levels leaves the building through cracks and pores, and outside air is drawn in through similar cracks in the facade at the upper stories.

Clearly the pressure gradient in the vertical direction is greatly affected by the distribution of openings in the façade. If we assume a uniform distribution throughout the exterior, then the outside pressure and inside pressure

Table 5.4 Equivalent Orifice Areas Equivalent Orifice Areas (square inches)	
13-inch porous brick wall, no plaster, 100 sq ft	3.1
Wall as above, 3 coat plaster, 100 sq ft	0.054
Frame wall, wood siding, 3 coat plaster, 100 sq ft	0.33
Door, tight fitting, 3 × 7 ft	7.6
Window, double-hung, loose fitting, 3 × 4 ft	4.7
Window, double-hung, tight fitting, 3 × 4 ft	0.93

will be equal at some midpoint. The height where this occurs is referred to as the "neutral zone."

Table 5.4 lists some equivalent orifice specifications for other building elements.[19] To illustrate the meaning of the information listed in Table 5.4, 100 sq ft of 13-inch porous brick wall (no plaster) has an area equivalent to 3.1 square inches open to the outside world. The stack effect, resulting from the temperature difference between the inside and outside air, is facilitated by openings in the building facade. The rate of air entering a specific equivalent orifice area as a result of the stack effect is calculated in the box titled Security Metric: The Vulnerability to Room Air Contamination Due to the Stack Effect.

The actual pressure across cracks and openings will be roughly the algebraic sum of the separate effects of pressure gradients caused by wind action, stack effect, and the building air supply and exhaust systems.[20] It is sometimes assumed that with the HVAC fan system on, buildings could maintain positive pressure relative to the outside minimizing the vulnerability to external contaminants. This is not necessarily so.

As a result of the stack effect, leaky structures would be susceptible to some contamination irrespective of the quality of mitigation implemented as part of the HVAC system since some of the air that enters the building will bypass the HVAC filters. The magnitude of the effect is highly dependent on the structural details of the building in question.

One way to minimize the stack effect might be to shut the heating or air conditioning off and close the vents but leave the HVAC fans running to provide internal air circulation. This would reduce the temperature gradient between the inside and outside environment. This does not happen instantaneously and the stack effect continues until the outside and inside air achieves temperature equilibrium. The engine for the stack effect is the temperature gradient between the internal and external air.

Another solution might be to use fans and filtering to overpressure/protect fire stairwells selectively pressurizing one compartment of the building. The filtering is essential since the air must be taken from somewhere to achieve an overpressure condition. These proposals should be vetted with a qualified building engineer before implementation.

SECURITY METRIC: THE VULNERABILITY TO ROOM AIR CONTAMINATION DUE TO THE STACK EFFECT

The pressure difference between the inside and outside of the building can be calculated from the following expression:

$$P = 7.6\ h\ (1/(t_0 + 460) - 1/(t_i + 460)) \qquad (5.21)$$

P represents the pressure difference between the inside and outside of the building (in units of inches of water; 250 Pascals (Pa) = one inch of water) as a result of the stack effect, h is the height above or below the neutral zone (in feet), t_0 is the outside air temperature (in °F), and t_i is the inside air temperature (in °F). It is clear from this expression that the location of the neutral zone is important in relation to air leakage since it determines the pressure difference at all other heights. The results of one study indicated that for large multi-zone buildings, the location of neutral zones under winter conditions is likely to be between one-third and two-thirds the height of the building.[20]

This expression results from the well-known fact that the density of air decreases with increasing height. The rate of decrease depends upon the temperature, and since the temperature inside and outside the building differs, the height-dependent pressure difference given by Equation 5.21 is the result.

Based on Equation 5.21 for P, a temperature difference of 20°F. between the inside and outside environments, and a height of 100 feet from the neutral zone will cause a pressure differential of 0.055 inches of water or 13.75 Pa (atmospheric pressure equals 14.7 lb/in^2 or 101,353 Pa). A 40°F temperature difference will cause a pressure differential of 0.115 inches of water (28.75 Pa) between the inside and outside environments.

Since there is a pressure gradient as a function of height above the ground within the building, outside air will seep into the building at different rates depending on the difference in pressure along the gradient. Various building structures have been characterized according to their leakiness and parameterized as having an effective aperture for a given surface area.[21] Assuming double-hung windows are the principal source of air leaks, the rate at which outside air fills a 1000 ft^3 room is calculated as follows:

The differential pressure due to the stack effect is given by the expression for P above. At h = 100 feet, t_o = 30°F, and t_i = 70°F, the differential pressure between outside and inside the building is 28.75 Pa or 0.115 inches of water.

(Continued)

SECURITY METRIC: THE VULNERABILITY TO ROOM AIR CONTAMINATION DUE TO THE STACK EFFECT—CONT'D

We want to know the rate at which air is entering the building as a result of the pressure difference induced by the stack effect. This is determined by the air velocity multiplied by the equivalent orifice area per room, A. Each room is assumed to have two 3 × 4 ft loosely fitting double-hung windows. According to the information in Table 5.4, such a window has an equivalent orifice area of 4.7 in^2, so each room has 2 × 4.7 in^2 = 9.4 in^2 = 0.066 ft^2 = A.

The pressure difference imparts kinetic energy to the air passing through the orifice area. The resulting velocity, v, of air caused by the pressure gradient is given by Bernoulli's conservation of energy equation which expresses pressure in terms of flow velocity. Rearranging terms in that equation yields the velocity

$$v = \sqrt{(2(p/\rho))}, \tag{5.22}$$

where p is the pressure in inches of water, and p is the density of air (at the specified temperature). In this case p = 0.115 inches of water, and the density of air at 30°F. is 0.081 lb/ft^3, so v = 22.46 ft/s.

V × A = flux of air into room due to the pressure gradient = 22.46 ft/s × 0.066 ft^2 = 1.48 ft^3/s.

Therefore, a 20 × 15 ft × 10 ft (i.e., 3000 cubic feet) room will be filled with outside air in just over half an hour. Under these conditions assuming there is no internal air pressure generated by the building HVAC system.

5.5.5 Vulnerability to chemical threats

Particulate filters can be effective in trapping biological and radiological contaminants but not in capturing vapors associated with chemical attacks except for tear gases and low volatility nerve agents such as VX, although a vapor component could exist for these agents.[22] In general, mitigating the effects of chemical agents requires the use of so-called sorbent filters.

Vapors released into the atmosphere could have an immediate and profound effect on individuals with whom these agents come in contact. In addition, vapors are likely to suffuse into the clothes of victims affecting others in contact with those originally impacted for some time after the initial attack. There are numerous chemical agents that would have a deleterious effect on individuals who inhale them, although chemical warfare agents such as phosgene and mustard gases are of particular concern.

Leaky buildings would be particularly vulnerable to external releases of chemical agents and the existing wind conditions would both disperse and dilute the vapors. As discussed previously, the temperature gradient-driven stack effect would play a prominent role in the vulnerability of a specific building to this attack. In winter months, contaminated air would enter through openings in the façade through lower floors and the reverse would be true in summer months as depicted in Figure 5.26. In this case as in the case of biological and chemical attacks, it seems reasonable to assume an attack would be initiated at ground level, but this is admittedly speculation on my part.

For wartime applications, increasing the density of delivery systems has been a principal goal of weapons designers since vapors so-modified would tend to linger on the battlefield amplifying their effect. Local atmospheric conditions play a significant role in the effectiveness of a chemical attack. Fortunately, an effective chemical attack is believed to be difficult to execute since proper dispersal is key to maximizing exposure before the agent evaporates or becomes too dilute.

Sorbent filters such as charcoal have been used in gas masks and are used to counter the "classic" chemical warfare agents. As noted previously, there are unfortunately a wide variety of potential chemical agents available and sorbent filters are not effective against all of these candidate weapons. Implementing sorbent filters could dramatically increase the load on an HVAC system, and as always, the risk versus gain associated with this strategy must be carefully weighed.

Additional information on sorbent filters and mitigating chemical attacks in general may be found in "Guidance for Filtration and Air-Cleaning Systems to Protect Building Environments from Airborne Chemical, Biological, or Radiological Attacks," US Department of Health and Human Services, National Institute for Occupational Safety and Health (NIOSH), Centers for Disease Control and Prevention, April 2003.

5.6 **THE VISUAL COMPROMISE OF INFORMATION**

Security professionals generally pay limited attention to the risk of an adversary merely observing visually exposed written material. This is evidenced by the fact that computer monitors, white boards, electronic presentations, etc., are often displayed in full view of windows that face buildings with unknown occupants. It is ironic that significant resources are expended by firms on sophisticated computer network monitoring as well as physical access technologies, and yet simple precautions to protect against the visual compromise of sensitive information displayed in the open are sometimes ignored.

Recall in our discussion on acoustics that we mentioned a phenomenon known as diffraction. This occurs when a wave of energy impacts an object and causes the energy to spread out thereby distorting the wave front. Diffraction also plays a very important role in optics. The ultimate resolution limit for a lens is set by diffraction and is proportional to the wavelength of the optical energy divided by the lens diameter. For circular lenses, the actual limit is 1.22 times the wavelength divided by the lens diameter.

A back-of-the-envelope calculation to assess the vulnerability to visual compromise of written information using commercially available optical elements suggests that the distance of vulnerability to visual compromise is about 100 m. I show the details by calculating the diffraction limit resolution of a lens (Equation 5.23) using reasonable values for operational parameters.

SECURITY METRIC: CALCULATING THE VULNERABILITY TO VISUAL COMPROMISES OF COMPUTER MONITOR DISPLAYS

Here we calculate the vulnerability to visual compromise in a scenario where a computer monitor is visible to a building directly across the street from your facility; i.e., where the observer and the vulnerable information are co-linear. The calculation for angular scenarios is more complicated. For a detailed examination of this problem see "Compromising Reflections or How to Read LCD Monitors Around the Corner" by M. Backes, M. Durmuth, and D. Unruh.

What is the distance of zero vulnerability with respect to the visual compromise of information displayed on a computer monitor by an adversary who is situated in front of a facility?

An adversary must be able to distinguish individual pixels on the computer screen to read the sensitive traffic. Let's assume this adversary is using a commercial telescope with a 300 mm diameter lens and he is directly facing the monitor of interest.

The diffraction-limited angular resolution θ for a circular lens is given by

$$\theta = 1.22\lambda/d, \tag{5.23}$$

where λ is the wavelength of visible light (assumed to be 5000 Å or 5000 $\times$ 10^{-8} cm), and d, the diameter of the lens, is 300 mm. θ represents the diffraction limit and is a physical limit imposed by nature. The above expression yields 2.03×10^{-6} radians (a measurement of angle; 2π radians is equivalent to 360 degrees) for the minimum angular resolution of the lens. The spot size of the image formed by the lens focused on the computer screen cannot exceed the size of an individual pixel to be effective since one must be able to distinguish individual pixels to read the text on the screen.

(Continued)

SECURITY METRIC: CALCULATING THE VULNERABILITY TO VISUAL COMPROMISES OF COMPUTER MONITOR DISPLAYS—CONT'D

If an individual pixel is 0.25 mm or 0.25×10^{-3} m in dimension, then the spot size created by the lens for a given distance cannot exceed 0.25 mm. The lower limit for the spot size of a focused beam as a function of distance from the source is set by

$$S = r\,\theta, \tag{5.24}$$

Where S is the spot size limit, r is the distance from the screen to the nosy bad guy, and θ is the minimum angular spread in radians as determined by diffraction. Solving for the distance r yields the maximum point for which a 0.25 mm spot size is achievable using this telescope. Therefore,

$$S = 0.25 \times 10^{-3}\,\text{m}$$
$$= r\ (\text{i.e., the distance from the adversary to the monitor})$$
$$\times\ (2.03 \times 10^{-6}) \tag{5.25}$$

Solving this equation for the distance yields r = approximately 100 m. This tells us that an adversary using a telescope with a 300 mm diffraction-limited lens could distinguish individual letters on the screen up to a distance of 100 m.

It is quite possible that our adversary might use a telescope with a larger lens; therefore the minimum angle of resolution θ will be proportionately smaller. On the other hand, the telescope might not have the highest quality lens and will not achieve the diffraction limit. Alternatively, a nosy observer might be at an off-angle with respect to the monitor (the distance will then be reduced by the cosine of the angle between the adversary and the front of the monitor).

However, and as I have steadfastly maintained throughout this book, analyses of security vulnerabilities are often exercises in approximation and are meant to bound the problem in the hopes of developing a practical mitigation strategy based on physically realistic conditions. In this case the analysis noted herein yields a ballpark estimate of vulnerability to the threat of visual compromise of information.

5.7 **SUMMARY**

Methods of estimating the vulnerability component of risk for a number of threats were discussed in this chapter. These included information loss from the unauthorized detection of electromagnetic and audio signal sources; explosives; computer network infections; biological, chemical, and radiological weapons; and the visual compromise of information. Simplifying physical assumptions have been made to develop ballpark estimates.

For some physical processes affecting risk, the vulnerability component of risk can be seen to scale according to a decreasing exponential as a function of scenario-dependent parameters, distance, and time. Another recurring theme is the inverse square with distance scaling of intensity for point sources of energy such as the sound intensity from a person speaking or the intensity of radiation emanating from a small chunk of radioisotope.

Vulnerability can be expressed using a number of metrics depending on the threat. In the case of the threat of information loss through unauthorized detection of electromagnetic or acoustic signals, it is possible to characterize this vulnerability in terms of the S/N ratio. In these cases the ambient noise in the frequency band of interest is a critical parameter in determining the threshold of vulnerability.

With respect to explosives, curves yielding the level of damage to structures in terms of explosive payload and explosive-to-target distance were shown. Scaling relations exist for explosive impulse and overpressure as a function of payload and the distance from the source. The effects of the impulse and overpressure are the physical quantities that cause damage to structures in an explosive incident.

A metric for the vulnerability to infections spread by e-mail on computer networks was postulated based on a scale-free distribution of linked nodes. In this theoretical formulation, the vulnerability to infection was related to the exponent that characterizes the distribution of linked nodes in the network. An explicit threshold parameter for infection growth was derived in terms of the rates of infection and remediation, and this threshold was proportional to network size.

Finally, the vulnerability to biological, chemical, and radiological threats was characterized in terms of the effectiveness of building HVAC systems as well as the leakiness of the exterior structure of facilities. Vulnerability to contamination by outside air was evaluated in terms of the specifications of HVAC particulate filtering as well as the room air exchange rate of the building interior with the external environment.

REFERENCES

1. Hall DE. *Basic Acoustics*. Harper and Row; 1987.
2. Ibid.
3. Ibid.

4. Bianchi C, Meloni A. Natural and Man-Made Terrestrial Electromagnetic Noise: An Outlook. *Annals of Geophysics.* 2007;50(3).

5. Ngo T, Mendis P, Gupta A, Ramsay T. Blast Loading and Blast Effects on Structures—an Overview. *EJSE Special Issue: Loading on Structures.* 2007;76–91.

6. Jeremic R, Bajic Z. An Approach to Determining the TNT Equivalent of High Explosives. *Scientific-Technical Review.* 2006;56:58–62.

7. *US Air Force Installation Force Protection Guide.* Guidehttp://tisp.org/index.cfm?cdid=10810&pid=10261.

8. Young C, Chang D. Probability Estimates of Explosive Overpressures and Impulses. 2008 (unpublished).

9. Pastor-Satorras R, Vespignani A. Epidemic Spreading in Scale-Free Networks. *Phys Rev Lett.* 2001;86:3200.

10. Chang D, Young C. Infection Dynamics on the Internet. *Computers and Security.* 2005;24(4).

11. Ebel H, Mielsch LI, Bornholdt S. Scale-free topology of e-mail networks. *Physical Review E.* 2002.

12. Chang D, Young C, Venables P. Complexity and Risk in a Network of Inter-related Computer Applications. 2007 (unpublished).

13. Lamarsh J. *Introduction to Nuclear Engineering.* 2nd ed. Addison Wesley; 1983.

14. Ibid.

15. Chang D. Vulnerability of Buildings to Radiological Dispersion Devices. 2008 (unpublished).

16. Garwin R. Private communication.

17. Brickner PW, et al. The Application of Ultraviolet Germicidal Irradiation to Control Transmission of Airborne Disease: Bioterrorism Countermeasures, *Public Health Rep.* 2003;118.

18. Wilson AG, Tamura GT. Stack Effect in Buildings. *Canadian Building Digest* (CBD-104).1968.

19. Wilson AG. Air Leakage in Buildings. Canadian Building Digest-23. *National Research Council of Canada,* 1961.

20. Ibid.

21. Ibid.

22. *Guidance for Filtration and Air-Cleaning Systems to Protect Building Environments from Airborne Chemical, Biological, or Radiological Attacks.* National Institute for Occupational Safety and Health (NIOSH), Centers for Disease Control; April 2003.

Mitigating security risk: reducing vulnerability

The best diplomat that I know is a fully-loaded phaser bank. — Lt. Cdr. Montgomery Scott

"A Taste of Armageddon," *Star Trek,* **Stardate 3192.1**

6.1 **INTRODUCTION**

The decision to invoke a particular risk mitigation strategy for a unique threat is in part dependent on the answers to five questions:

1. What is the likelihood or potential for threat occurrence?
2. What is the vulnerability to loss assuming a threat does occur?
3. Do the consequences of a threat merit mitigation (i.e., what will be the impact of an occurrence)?
4. What methods of risk mitigation are available?
5. Can my company afford the required mitigation and/or are there less expensive options to manage the risk associated with a threat?

There is often more than one way to mitigate the risk associated with a particular threat. In the case of threats characterized by natural phenomena it may be possible to determine the effectiveness of risk mitigation with relative precision. Assuming the relevant physical quantities and the scaling of scenario-dependent parameters are understood, this should enhance confidence in the resulting mitigation strategy.

However, this is not always possible. There are threats that are either divorced from the physical world and/or do not lend themselves to analysis via quantitative methods. At times the only available recourse is to develop an intuition-based mitigation strategy or merely live with the threat, be vigilant, and manage expectations.

In this chapter we discuss various measures to mitigate the risk associated with some security threats for which there is growing concern. It is unclear if these "nontraditional" threats are actually more likely to occur. However, in the wake of 9/11 the specter of the mere potential for occurrence has heightened public awareness and fear. In addition, mitigation of the threat of information loss due to the unauthorized detection of electromagnetic and acoustic signal energy (i.e., speech) is discussed. Quantitative background material is also presented that will help facilitate the development of metrics that can be used to estimate the effectiveness of mitigation measures.

6.2 **AUDIBLE SIGNALS**

As discussed in Chapter 5, information loss is a concern in today's information-intensive business environment. You have probably noticed that in this book considerable attention has been focused on this topic, and I have specifically examined the vulnerability to unauthorized detection of acoustic (mechanical) and radio frequency/optical (electromagnetic) signal energy.

It turns out that the vulnerability to signal detection can be quite varied depending on the presence of reflecting and/or absorbing surfaces along the energy propagation paths. Although little attention is given to the problem of information loss due to unintended overhears of audible conversations, and relevant statistics on such losses are not readily available, reducing this risk from within facilities such as conference rooms as well as public areas arguably deserves more attention in the security community.

In the case of speech, the intensity of energy propagating in open areas decreases according to the now familiar $1/4\pi r^2$ expression where r is the distance from the source to the listener. In enclosed areas and as noted previously, some of the energy will repeatedly be reflected by the higher density walls rather than propagating away from the source and dissipated by friction. This process could actually contribute to signal amplification.

Let's consider the outdoor audibility scenario first. We are concerned about the threat of unauthorized individuals overhearing sensitive conversations discussed in public areas. Of course there is not much that can be done for employees who behave irresponsibly by loudly discussing sensitive information near unknown and/or unauthorized listeners. However, one could conceive of an outdoor area accessible to employees contiguous with space regularly accessed by non-employees or even competitors. The question then is how to reduce the vulnerability to unauthorized overhears in such a scenario.

There seems to be four fairly obvious choices: (1) reduce the source intensity (i.e., speak more softly), (2) increase the distance between the source and the unauthorized listener, (3) increase the background noise, or (4) install a barrier between the source and unauthorized listeners.

Speaking sotto voce sounds like a simple solution, but people tend to forget and gradually lapse into their natural speaking voice. In addition, even if the exact threshold intensity was calculated correctly and was translated into the proper speaking level, continuously enforcing obedience would be a daunting task.

Increasing the distance between source and listener could definitely work assuming this is feasible. Companies usually have a limited amount of available real estate and often do not have the luxury of expanding their perimeter. One could restrict areas as off-limits and thereby add a buffer zone. In this instance a calculation as described in Section 6.2.1 is required to estimate the vulnerability component of risk. As always, due consideration must be given to the ambient noise across the audible frequency spectrum.

As an aside, if a competitor is serious about overhearing conversations and decides to mount a more sophisticated attack, that competitor might not simply rely on the human ear as a detector. He could deploy a parabolic antenna to provide signal gain assuming there was an unobstructed air path between the speaker and the microphone. In this case the vulnerability previously calculated using the naked ear is increased by a factor of $k(\pi D)^2/\lambda^2$, where D is the radius of the parabolic dish, λ is the wavelength (the wavelength of a 1 kHz audio tone in air is about 1 foot), and k is an efficiency factor of about 0.55. For a 1-foot diameter dish, this yields a gain of about 1.7 above the sound intensity detected by the ear alone at 1 kHz.

Raising the background interference in an unsophisticated way such as using a musical recording is not going to help the situation against a slightly sophisticated adversary except possibly to interfere with employee conversations. Ironically, this could make matters worse by causing the employees to raise their voices above the level of the interference and enhance an adversary's signal-to-noise (S/N) ratio.

In the same vein, if the interfering source is not physically close to the audible signal you hope to protect, a properly aimed directional microphone will intensify the signal of interest at the expense of the interfering sound. Achieving processing gain for signals in the presence of non-random and spatially localized sources of interference is relatively easy.

6.2.1 **Acoustic barriers**

In some circumstances, it might be feasible to erect a barrier in an effort to reduce the vulnerability to information loss by unauthorized listeners. What specifications for barriers are necessary to mitigate the vulnerability component of risk for this threat?

I briefly mentioned the concept of diffraction in Chapter 5. This is a process where objects that have similar physical dimensions as the wavelength of the interacting energy cause that energy to spread out in space. We can use the principle of diffraction to estimate the attenuation of sound energy caused by the emplacement of a boundary between a sound source and potential listener at a given frequency.

The theory of diffraction is a complex subject and is beyond the scope of this book. However, simple geometry can be used to calculate the effect caused by a barrier in line with an audio source.[1] Such an analysis uses the Fresnel (pronounced FrenNEL; this is the same fellow for whom lenses on photocopiers and in lighthouses are named) number: the number of half-wavelengths by which the shortest path of the audio energy grazing the barrier exceeds the straight line path through the barrier.

Specifically, the Fresnel number equals

$$N_f = (A + B - (L + d))/(\lambda/2), \tag{6.1}$$

where L is the distance of the source to the wall, d is the distance of the receiver to the wall, A is the distance from the source to the top of the wall, B is the distance from the receiver to the top of the wall, and λ is wavelength (see Figure 6.1).

The insertion loss in units of decibels is defined in terms of the Fresnel number N_f as follows:

$$\text{Insertion Loss (IL)} = 16 + 10 \log N_f. \tag{6.2}$$

The expression for insertion loss indicates how much the sound level is reduced due to the presence of the barrier. Note the presence of wavelength in the denominator of the expression for N_f. This means that for large wavelengths the insertion loss is reduced relative to a fixed barrier height. This fact has explicit risk implications.

In other words, longer wavelengths/lower frequencies are more difficult to attenuate than shorter wavelengths/higher frequencies. In general information-carrying signals are comprised of many wavelengths/frequencies. Physical effects often vary with frequency/wavelength and an analysis of signal behavior must be performed in several regions of the spectrum to give a complete appreciation of the risk of signal detection. In this case

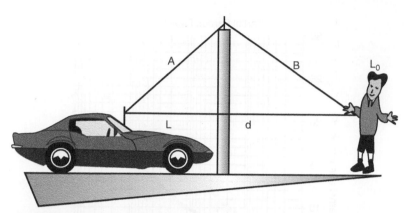

■ **FIGURE 6.1** Geometry of acoustic insertion loss. *From http://www.engineeringtoolbox.com/outdoor-sound-partial-barriers-d_65.html.*

the goal is to quantify the attenuation achieved by using a barrier to block sound. This involves calculating the insertion loss produced by the barrier for representative frequencies of audible sound energy.

Using a worked example from the book *Basic Acoustics*,[2] for a barrier of height 4 m, a source located 6 m from the wall, and a receiver located 2 m from the wall, A equals 6.32 m, B equals 8.55 m, and L + d is 14.04 m, at a frequency of 1.7 kHz (wavelength of 0.2 m). By geometric arguments, the Fresnel number is calculated to be 8.3, and the insertion loss is 25 dB.

What happens at 10 times the wavelength that corresponds to 1/10 the frequency or 170 Hz? In this case $\lambda/2$ is 1 m so the Fresnel number is 0.83 and the associated insertion loss is calculated to be 15 dB. An increase in insertion loss would require that the quantity A + B − (L + d) in the expression for N_f be proportionally increased. This means the barrier height would have to be raised illustrating the difficulty in blocking low frequency audio signals using barriers.

Sound attenuation by a barrier as a function of the Fresnel number is shown in Figure 6.2.[3]

With respect to barrier placement, it is best to position the barrier as near to the source as possible for maximum sound attenuation. This causes the propagating wave to travel the maximum distance through the air after the barrier and to negotiate the steepest angle between the source and listener. The second best position is near the receiver/listener and the worst position is midway between the two.

In Chapter 5 I discussed the effect of walls in evaluating the vulnerability to unintentional overhears by a listener in a contiguous room where a

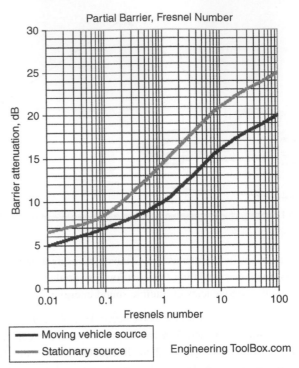

■ **FIGURE 6.2** Sound attenuation by physical barriers. *From http://www.engineeringtoolbox.com/outdoor-sound-partial-barriers-d_65.html.*

conversation was occurring. In that discussion I inadvertantly touched on risk mitigation since sound attenuation across the barrier was calculated. We observed how adding mass to barriers attenuates sound energy and that attenuation scales linearly with increasing frequency in some frequency regimes.

For low frequency acoustic signals we must do something such as adding mass to a barrier to mitigate the risk of detection by an unauthorized listener. This could quickly become impractical unless we have an extremely hefty foundation to support the added weight. Fortunately there is a better way.

6.2.2 **Sound reflection**

Recall the concept of acoustic impedance mismatch introduced in Chapter 5. This is relevant when energy such as sound propagates through materials of different densities. The propagation of energy is highly dependent on the medium in which it travels. The square of the reflection coefficient, R, represents the fraction of propagating audible signal power that is reflected and equals the square of the difference of acoustic impedances (i.e., the product of density and velocity $= \rho v$) divided by the sum of impedances:

$$R^2 = ((\rho_1 v_1 - \rho_2 v_2)/(\rho_1 v_1 + \rho_2 v_2))^2, \tag{6.3}$$

Here ρ_1 and ρ_2 could for example, represent the densities of wood and air, respectively, and v_1 and v_2 are the velocity of sound in wood and air, respectively. The fraction of signal power not reflected or absorbed by the material is the power actually transmitted across the barrier and therefore available to a nosy listener. Recall that Appendix C lists the speed of sound at room temperature for various materials.

The reflection coefficient results from the acoustic impedance mismatch from sound traveling between materials of different densities. Therefore the sound energy impinging on the higher density wall via the lower density air produces reflected energy and serves to partially mitigate the risk of listeners who are up to no good.

It turns out that plugging in some numbers into the expression for R^2 reveals that about 99% of the sound energy is reflected back toward the source at the air–wood interface. However, this sounds better than it is literally and figuratively, as the situation does not guarantee immunity from detection by a nosy listener. As noted previously, some of the reflected acoustic energy from the originating room leaks across the barrier via multiple reflections of the acoustic energy wave at the air–solid interface. Therefore, each reflection will cause 1% of the acoustic energy to leak across the barrier underscoring the role of reverberation in audible scenarios.

In addition, sound energy traveling in dense media travels with greater speed and propagates further than in less dense media before being attenuated. This is cause for concern in buildings with multiple tenants. It is also a building feature that should be investigated by technical surveillance countermeasure professionals as noted in Section 6.10.

A practical solution to both the problem of leakage of energy across the boundary between the two rooms and the propagation of acoustic energy within building structural elements is to add sound-absorbing material to elements exposed to speech-borne energy.

6.2.3 **Sound absorption**

Sound energy is absorbed by virtue of its conversion to another form of energy, principally heat. Heat is generated due to the contact between vibrating air molecules (recall that these vibrating molecules are sound energy) and the resistance to excitation. This process is more compactly described as friction, which results in the generation of heat albeit a very small amount given the low levels of power in speech-borne acoustic energy. The following is meant as a high level overview distilled from a useful Web site on sound.[4]

There are different types of acoustic absorbing material that can be applied to a surface to reduce the fraction of transmitted energy. Porous absorbers such as mineral wool, fiberboard, or plastic foams have an open pore structure. The vibrating air molecules are forced through the pores, attenuating the vibrations, and a conversion to heat occurs in the process. High frequency (or equivalently short wavelength) acoustic energy is preferentially absorbed by porous materials.

If the thickness of the absorbing material is much less than a ¼ wavelength of the incident acoustic energy, then the effectiveness of this method is reduced since the acoustic pressure wave reaches its maximum amplitude at ¼ of a wavelength. Displacing the absorbing material between ¼ and ¾ wavelengths from the wall will increase the effectiveness of this method and sometimes obviates the need for thicker absorbing material.

Membrane absorbers can exist in the form of flexible sheets stretched over supports or as rigid panels displaced from the front of a solid wall separating the two rooms. Mounting the material in this way produces an air gap between the absorber and the barrier wall. Absorption occurs as a result of the conversion of the air vibrations to heat based on the resistance of the sheet to rapid flexing and the resistance of the enclosed air to compression. These types of absorbers are most effective at the resonant frequency determined by the surface density of the panel $= M$ (kg/m^2) and the width or depth of the air gap $= b$ (meters), where the resonant frequency is given by F (Hz) $= 60/(M*b)^{1/2}$.

Recalling the mass on a spring depicted in Figure 5.4, the resonant frequency of this system is larger because the effective spring constant is increased; the compressed air in the air gap behind the membrane adds to the restoring bending force. Also, the frequency can be increased by using thin membranes since that decreases the value of the surface density, M. Membrane absorbers are most effective at low frequencies.

Cavity absorbers are air containers with a narrow neck. The air in the cavity acts like a spring at the resonant frequency of the enclosed air volume. These types of absorbers are highly absorbent, but only in a very narrow range of acoustic frequencies centered on the resonant frequency (i.e., the frequency of maximum absorption). The resonant frequency, F, is given by $(340/2\pi) \times (S/VL)^{1/2}$, where S is the cross-sectional area of the neck (m^2), V is the volume of the cavity (m^3), and L is the length of the neck (meters).

Perforated panel absorbers combine all three methods in one system. The panel may be composed of plywood, hardboard, plasterboard, or metal, and may also act as a membrane absorber. The perforations in the panel function as cavity resonators augmented by porous absorbing material. Perforated panel absorbers constitute the bulk of the broad frequency commercial acoustic insulating materials.

Interposing some form of air barrier is a good means of reducing vulnerability to unauthorized detection of audible information. Adding even a thin layer of air between walls or barriers separating rooms can yield significant results. One note of caution: any structural elements or conduits such as studs and/or ducts that physically connect rooms of concern will offer pathways for sound conduction and must be addressed by using appropriately insulating materials.

Finally, a security metric is sometimes used in the building industry to specify sound attenuation by building materials such as wall board. This is known as the Rw value, and is a single number (denoted in units of decibels of course!) that accounts for sound attenuation across the audible frequency spectrum (see Appendix H).

6.3 **ELECTROMAGNETIC SIGNALS**

Electromagnetic signals represent the second of two big groups of information-carrying signal energy. Although electromagnetic signals conduct energy via wave-like motion, their interaction with materials can be quite different from mechanical energy. However, and as the case with sound, the effects of intervening material on electromagnetic signals are very important in understanding vulnerability to unauthorized signal detection.

Because of the presence of reflecting and absorbing surfaces, predicting the amplitude of electromagnetic signals in environments other than open spaces is difficult. This is especially problematic in complex urban environments. Nevertheless, the theoretical equations governing electromagnetic interaction with matter are well understood. In addition, experimental measurements of the propagation of signals in various environments have been made that can provide approximate guides for mitigating the vulnerability component of risk.

6.3.1 **Electromagnetic shielding**

What about the effectiveness of various materials in attenuating information-carrying electromagnetic signals? Can we use materials for shielding to reduce the vulnerability to signal detection and potential information loss? For those of us who listen to commercial radio broadcasts (or use a cell phone or any other wireless electronic device), we know from first-hand experience that air does little to stop electromagnetic energy but atmospheric conditions can definitely play a role in radio signal propagation.

There are a number of important physical parameters that influence the propagation of electromagnetic energy through materials. Most notably these include the frequency of the wave and the electrical conductivity of the material with which it interacts. Electromagnetic waves are relatively unaffected by some of the more common building materials such as wood, drywall, and glass, unless the energy is in the upper portions of the microwave spectrum. Concrete with rebar can definitely affect radio frequency transmissions and the thickness of slabs can be important in determining floor-to-floor transmission losses.

Table 6.1 provides the measured attenuation of a 5.8 GHz electromagnetic signal when propagating through a variety of building materials.[5] Results are shown for both parallel and perpendicular polarizations of the electromagnetic wave, which means that the attenuation may differ if the orientation of the oscillating electric field is parallel to the barrier or perpendicular to it.

Attenuation results vary considerably for different frequencies and 5.8 GHz is a relatively high frequency in the spectrum of commercial electromagnetic signals. In general, higher frequencies suffer more significant attenuation through materials than lower frequencies. This topic has been

Table 6.1 Electromagnetic Energy Attenuation Due to Common Building Materials (5.8 GHz)

Building Material	Parallel Polarization Attenuation (dB)	Perpendicular Polarization Attenuation (dB)
PVC	0.4	0.6
Gypsum plate	0.8	0.7
Plywood	0.9	0.9
Gypsum wall	1.2	3.0
Chipboard	1.3	1.0
Veneer board	2.2	2.0
Glass plate	3.2	2.5
6.2-cm soundproof door	3.4	3.6
Double-glazed window	6.9	11.7
Concrete block wall	11.7	9.9

extensively studied and numerous measurements of electromagnetic attenuation by building materials have been conducted in other frequency regimes.[6]

For ballpark estimates, an average of the polarizations should suffice for attenuation calculations. Finally, recall from our discussion in Chapter 2 that the calculation of attenuation from multiple layers of material can be accomplished by directly adding the decibels of attenuation for each interposing layer.

High conductivity materials such as metals have exceptionally good shielding properties, although the quality of shielding can vary even among metals. How does electromagnetic shielding actually work?

As the name implies, all electromagnetic signals have both electric and magnetic field components that oscillate in mutually perpendicular planes and are 90 degrees out of phase. A useful figure that illustrates electric and magnetic field oscillation can be found at http://www.casde. unl.edu/tutorial/rs-intro/images/wavedia.gif. Electromagnetic shielding involves the attenuation of both components. The wave or radiation impedance of an electromagnetic wave is defined as the ratio of its electric to magnetic component in the media in which it is propagating, and this ratio varies depending on the source of the signal and the distance from that source.

If the intrinsic electromagnetic impedance (impedance in this case can be thought of as frequency-dependent resistance) of the wave in air differs from its value in metal then the incident wave energy is reflected. The greater the difference in impedance of the incident wave and the intrinsic impedance of the material it encounters, the more energy is reflected at the boundary between the two materials. For the engineers out there, the wave impedance in air is purely resistive whereas the intrinsic impedance of metals includes a significant reactive term.[7]

Since the intrinsic electromagnetic impedance of metals differs considerably from air, this mismatch causes most of the electromagnetic wave energy to be reflected at an air–metal interface when the distance between the interface and the source is much larger than the energy wavelength.

However, if the metal is closer to the electromagnetic source than a wavelength the situation can change. For example, if the source is a current loop, then near the source the electric field is very small, i.e., the impedance is very small when the magnetic field dominates the electric field close to the source. Most metals have intrinsic impedance values of milliohms, and this value is closer to the lower wave impedance. Consequently, magnetic field energy close to a loop is reflected less than electric field energy and magnetic fields are more difficult to shield, especially at low frequencies.

For electric fields at distances that are short compared to a wavelength of the incident energy and are generated by a source that creates primarily electric rather than magnetic fields near the source (e.g., a short, straight wire), the impedances are high relative to the intrinsic impedance of the metal and the incident wave undergoes significant reflection upon contact with a metal surface.

In analogy with acoustic energy, the incident electromagnetic energy is either reflected, absorbed, or transmitted by a barrier. According to Maxwell's equations, the field also induces a current in metals, and this residual current generates its own magnetic field. The current density at any depth in the shield and the rate of current decay as a function of shield thickness is governed by the conductivity of the metal and the magnetic permeability of the metal, as well as the frequency and amplitude of the field source. Figures 6.3[8] and 6.4[9] illustrate the shielding mechanism.

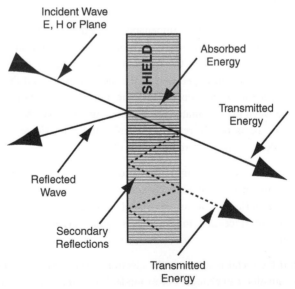

■ **FIGURE 6.3** Attenuation of Electromagnetic Interference (EMI) by a Shield. Shielding of electromagnetic signals. *From www.chomerics.com. With permission.*

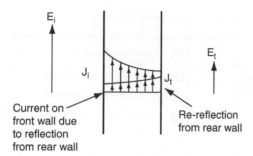

■ **FIGURE 6.4** Electric fields and currents in electromagnetic shielding. Variation of current density with thickness for electrically thin wall. E = electric field strength; J = current density; I = initial; and t = transmitted. *From www.chomerics.com. With permission.*

Electromagnetic fields do penetrate the surface of metals but not to great depths. The so-called "skin depth" is a physical characteristic of metals that specifies this penetration distance, and is inversely related to the square root of the product of the metal's conductivity and the wave frequency. Therefore, for a given metal conductivity, higher frequency electromagnetic signals penetrate metals less than those of lower frequency. Typical skin depths in metals are on the order of microns.

Truly effective electromagnetic shielding is not easy to achieve and if done incorrectly can make matters worse. Reflections cause the electromagnetic signal to bounce around and these bounces can add or subtract at a given point in space, sometimes resulting in signal enhancement or gain.

Figure 6.5[10] illustrates the shielding effectiveness (i.e., the attenuation in dB) of copper as a function of frequency for electric fields, magnetic fields, and electromagnetic fields (i.e., plane waves or fields distant from the source compared to the radiation wavelength). This is a somewhat complicated graph, but notice that the reflection losses are extremely high in copper for electric fields and plane waves below 10 MHz and are minimal for magnetic fields in the same frequency regime. Absorption losses begin to dominate for metal thicknesses on the order of thousandths of an inch above 10 MHz for magnetic fields and copper.

6.3.2 Intra-building electromagnetic signal propagation

Reliable indoor propagation of radio signals is an important factor in operating wireless electronic communication devices. Remember that security professionals are in some sense at odds with network providers

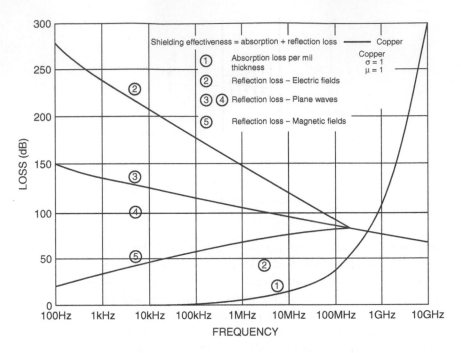

Shielding Effectiveness of Metal Barriers

■ **FIGURE 6.5** Shielding effectiveness of metal barriers. *From www.chomerics.com. With permission.*

in terms of their respective goals. The former typically seek to contain such signals and reduce the probability of unauthorized signal detection, whereas network providers aim to maximize reliability and coverage. The problem with accurately characterizing this problem from a security perspective is that the results are highly scenario-dependent, which will be evident from the following discussion.

People have measured the attenuation of signals in buildings and have established ranges of values that can be used as a guide to evaluating vulnerability to unauthorized signal detection and resultant information loss. One should always keep in mind that these figures are approximate when assessing the vulnerability component of risk. It is useful to understand the general behavior of indoor and outdoor electromagnetic propagation to assess the effectiveness of mitigation strategies.

Measurements of propagation loss in "typical" office buildings at 914 MHz (i.e., the cell phone frequency band) reveal losses of between 50 and 90 dB for a 10 m separation.[11] This corresponds to variations in loss ranging between factors of a hundred thousand and a billion. A loss factor of a billion may sound like a lot but radio receivers can detect very low power signals ($\sim 10^{-12}$ W). Engineers refer to receivers with great signal sensitivity as having a wide dynamic range.

As I have stressed repeatedly, ambient noise is a big factor in the vulnerability to signal detection and this applies to security professionals and adversaries alike. Losses will be a factor of 4 or so higher at 2.4 GHz (i.e., a frequency of 802.11 wireless LANs) and will be reduced by 3 dB per octave or each doubling of frequency thereafter. Wireless systems have typical communication "link budgets" of about 120 dB and most of the signal is attenuated in the first 10 m of propagation.

This is all good news from a security perspective. Note that these data include inter- and intra-floor propagation paths. Concrete slabs of buildings cause significant attenuation depending on the transmit frequency and a 10 m propagation path in the vertical direction will comprise about 3 stories or at least 3 concrete slabs. Concrete and steel flooring allows for better propagation than all-steel flooring. These figures can serve as a guide in estimating vulnerability to signal detection within multi-story (and often multi-tenant) buildings.

The communication picture improves (or worsens from a security risk perspective) for scenarios of same-floor propagation. Such measurements indicate 60 dB of attenuation at a 10-m separation distance. At 50 m, measurements reveal 110 dB of signal attenuation. Therefore, nearly the entire communication budget of 120 dB is "eaten up" at a distance of 50 m for same-floor propagation.

One should be aware that the situation differs at the higher frequency of 2.4 GHz.

For scenarios that include inter- and intra-floor communication links, the intensity of transmitted signals propagating indoors at this frequency has been found to scale with distance as the inverse 3.5 power (i.e., $1/r^{3.5}$ where r is distance from the radio frequency source). Therefore, the intensity of signals radiating indoors attenuates much more rapidly with distance from the source than for point source transmitters operating in open spaces.

The expression for path loss where the scaling relationship of $1/r^{3.5}$ for indoor signal propagation loss applies is given by:

$$\text{Path Loss (dB)} = 40 + 35 \log r \text{ (meters)} \tag{6.4}$$

(Note: $\log_{3.5} X = 35 \log_{10} X$)

So a 10-m indoor separation between source and receiver will yield a 75-dB path loss. At 100 m, the system will experience $40 + 35 (\log 100) = 110$ dB of path loss. Using the system link budget figure of 120 dB, an indoor communication link at 2.4 GHz is at the limit of performance at about 100 m. It should be noted that these results cite scenario-dependent variances in measurements of about 13 dB.

Other cited research has yielded different results for indoor propagation of electromagnetic signals and incorporates building elements in this model.[12] Specifically, an expression specifying path loss for indoor propagation is given by

$$30 \text{ dB (i.e., the power loss at 1 m)} + 10 \text{ n } \log (d) = kF + LW \qquad (6.5)$$

Here n = a power-delay index, d = distance between transmitter and receiver, k = number of floors the signal traverses, F = loss/floor, L = number of walls the signal traverses, and W = loss/wall. Clearly the presence of floors and walls plays a major role in intra-building electromagnetic propagation. In this study n varied between 1.6 and 3.3 depending on the type of facility with significant variations in model accuracy for typical offices and stores.

This study also found that signal strength depended on the existence of open plan offices, construction materials, density of personnel, furniture, etc. Path loss exponents varied between 2 and 6 (i.e., between $1/r^2$ and $1/r^6$... quite a variation!), wall losses were shown to be between 10 and 15 dB, and floor losses varied between 12 and 27 dB. The lesson here is that any such calculation should be used principally to bound the problem and develop a physically realistic mitigation strategy.

6.3.3 Inter-building electromagnetic signal propagation

Security scenarios of interest sometimes include the vulnerability to unauthorized detection of inter-building communications, i.e., a commercial communication device such as a wireless LAN or wireless headset is transmitting from within one building and the concern is that an adversary can detect this signal from within another building in proximity.

Again, the bad news for assessing building losses for this radio frequency propagation scenario is the high degree of variability in reported measurements.[13] Such losses depend on building materials, orientation, floor layout, building height, percentage of windows, and the transmission frequency. It has also been observed that (1) the received signal strength increases with increasing height within the building since the urban radio frequency clutter/noise is reduced with increasing elevation; (2) the penetration loss decreases with increasing frequency, presumably because higher frequencies correspond to shorter wavelengths that are less affected by diffraction relative to the size of apertures such as windows, cracks, etc.; and (3) the loss through windows is typically 6 dB (i.e., a factor of 4) or less.

For outdoor-to-indoor propagation scenarios, the scaling relationship of signal intensity with distance has been measured to be $1/r^{4.5}$ on average, and ranges between $1/r^3$ to $1/r^{6.2}$.[14] Try plugging in a few values for r to get a feeling for the difference even a small exponent can make in the rate of decrease in intensity with increasing distance. For example, moving the distance between source and receiver by a factor of 10 (say from 10 to 100 m) causes the intensity of a $1/r^3$ falloff to be reduced by a factor of 1000. For the same change in distance, a scaling relationship of $1/r^{6.2}$ will cause an intensity reduction of nearly 1.6 million.

Attenuation by the building structure has been measured to be between 2 and 38 dB. This also represents a huge variation and illustrates the difficulty in making accurate estimates of vulnerability to radio frequency detection for inter-building scenarios. The best situation would be to make an actual measurement of loss in the scenario of interest, but this is often not feasible. So prudence demands assuming lower levels of attenuation when developing a mitigation strategy.

The aforementioned discussion relates to what we might refer to as "natural" mitigation effects; i.e., the vulnerability to unauthorized signal detection and potential information loss for electromagnetic signals is helped by natural phenomena that absorb, diffract, reflect, and generally interfere with signal propagation. This does not in any way suggest a general immunity from unauthorized detection since commercial devices are designed to be robust. However, physical effects as described herein generally play an important role in estimating the risk of detection for specific scenarios.

6.3.4 **Non-point source electromagnetic radiation**

Thus far mostly point sources of electromagnetic radiation have been considered. As we have now noted many times, the intensity of energy emanating from point sources scales as $1/r^2$ in an ever-expanding spherical pattern of decreasing intensity at increasing radii from the source. What about sources/devices that concentrate their radiated energy within a volume of space rather than transmitting energy equally in all directions?

Dish antennae, antenna arrays, and lasers are good examples of such sources. How vulnerable are these to interception based on their limited physical footprint? Stated differently, what mitigation benefits are derived from limiting the spatial presence of the signal to reduce its vulnerability to unauthorized signal detection?

I will begin by indicating what it means to focus energy. If we start with an initial amount of energy derived from a source such as an electronic transmitter, a speaker of audible information, light rays, etc., and do not interfere with the signal, we cannot physically add to the original amount of transmitted energy. What you start with is what you get in terms of total energy radiated. However, we can affect the *density* of energy as it propagates through space by limiting the volume through which it propagates. Focusing effectively confines the energy to a smaller element of space, but the total radiated energy remains the same.

The ability to focus electromagnetic energy is limited by diffraction and I indicated in Chapter 5 that for circular apertures the minimum angular spread or divergence of electromagnetic energy is given by the expression $1.22 \, \lambda/d$ where λ is the energy wavelength and d is the diameter of the lens or antenna. We see from this expression that the limit on the width of the beam (i.e., the limit on focusing) increases when either the wavelength is smaller or the aperture is increased. The bigger the lens or antenna the narrower or more focused the beam for a given wavelength.

We also learned that antenna gain is a measure of the signal enhancement, and that for a dish antenna this enhancement equals $k\pi A/\lambda^2$ where k is an efficiency factor (about 0.55), A is the area of the mouth of the dish, and λ is wavelength.

What about the use of short wavelength devices such as optical lasers to transmit signals? From the expression for the diffraction limit we would expect the angular divergence from a device transmitting a very short wavelength signal to be narrow. That is precisely the case with an infrared laser where the wavelengths are on the order of a micron and largely invisible to the naked eye. In the box titled Security Metric: The Probability of Fortuitous Detection of a Laser Communication Signal, the probability of randomly or accidentally detecting a laser signal is calculated. It turns out that this is not likely due to limited beam divergence.

Radio frequency transmitters focus electromagnetic energy by using antennae that produce signal gain. This is good for both improving the S/N ratio as well as in mitigating the threat of signal detection and attendant information loss. However, the operational problem in implementing this type of mitigation is that the transmitter and receiver must be properly aligned to ensure reception. The more gain provided by the lens or

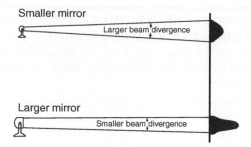

■ **FIGURE 6.6** The effect of aperture size due to diffraction.

antenna, the greater is this alignment criterion since the transmitted signal is confined to narrower slice of space.

Figure 6.6 illustrates the concept of beam divergence from two different diameter apertures, in this case mirrors. As previously noted, a larger aperture produces a smaller divergence or more tightly focused beam due to the effects of diffraction.

SECURITY METRIC: THE PROBABILITY OF FORTUITOUS DETECTION OF A LASER COMMUNICATION SIGNAL

In this case we are interested in the level of privacy afforded by a high gain (i.e., spatially confined) signal relative to "random" (i.e., unintentional) detection by an unauthorized individual. A laser has a relatively narrow beam divergence and I assume this is 0.3 mrad derived from technical specifications of commercial lasers.

A radian is a measure of angle and 360 degrees corresponds to 2π radians. Based on this divergence figure, the diameter or spot size of the laser beam at a distance of 100 m will be $0.3 \times 10^{-3} \times 100$ m $= 0.03$ m $= 3$ cm. We want to know the likelihood of detecting the laser signal by someone using a 300 mm telescope lens at a distance of 100 m from the source in any direction but in a fixed plane. I have made a significant assumption that the laser and detector are vertically aligned. In real life this would further decrease the probability of fortuitous detection.

Let the circumference of a 100 m radius circle $= C$, the diameter of the telescope lens $= Y$, and the divergence of the laser beam $= X$. The probability that the laser beam falls entirely within the lens equals the probability of the lens being in the proper segment on the circumference of the circle times the probability that the beam falls entirely within that segment. This is given by $(Y/C) \times (X/Y) = X/C$.

The beam divergence at 100 m was calculated to be 3 cm or 0.03 m and the circumference of a circle of radius 100 m is $2\pi \times 100$ m $= 628.32$ m.

(Continued)

SECURITY METRIC: THE PROBABILITY OF FORTUITOUS DETECTION OF A LASER COMMUNICATION SIGNAL—CONT'D

Therefore, $X/C = 0.03/628.32 = 4.77 \times 10^{-5}$ = probability of randomly detecting a laser beam using a telescope with this diameter lens at 100 m from the laser source.

Now compare this result with the probability of privacy using a 10 cm wavelength microwave beam ($\sim$19 GHz) and assuming one meter diameter transmit dish. We find the diffraction limited beam divergence is 0.12 radians.

At a distance of 100 m the beam is 12 m in diameter which in this case is much wider than the beam. If X = the beam diameter, Y = the receiver dish diameter = 1 meter and C is the circumference of the 100 m radius circle, the probability of fortuitous detection is $X/C \times Y/X = Y/C = 1/628 = 1.6 \times 10^{-3}$.

6.4 VEHICLE-BORNE EXPLOSIVE THREATS: BARRIERS AND BOLLARDS

The threat of the delivery and detonation of explosives near or in buildings can represent a significant threat. A study of historical data yields abundant information on the use of this technique by terrorist groups. Conventional explosives are relatively easy to acquire, and a small amount can go a long way with respect to destructive capability.

In Chapter 5 we observed how explosive effects are affected by distance between the explosive source and the target. Approximate scaling relationships were specified as $1/r^3$ with respect to overpressure and $1/r$ for impulse where r is the distance between the explosive source and a target structure. It is important to note that overpressure and impulse are functions of both explosive payload and distance. The situation gets more complicated in spaces crowded with large structures, but these scaling relationships can serve as a decent approximation to reality in open spaces. Recognizing that standoff distance is critical to protecting physical assets from this threat, the use of a control in the form of physical barriers, and the implementation method of bollards in particular, has increased dramatically in recent years.

Bollards are designed to prevent vehicles from breaching a prescribed boundary and detonating their payload too close to or within building structural elements. Bollards are rated according to their ability to absorb the kinetic energy of a moving vehicle, and this energy rating varies based on the bollard type and installation details. Developing a mitigation strategy with respect to the threat of a fast approaching vehicle packed with explosives will be informed by an understanding of bollard performance specifications relative to vehicle weight and speed. Specifically, the goal is to ensure attacking vehicles cannot get too close to their intended target as determined by a rigorous assessment of the vulnerability component of risk (see Section 5.3).

I will assume that readers have done their homework and established the required separation distance of the bollards from the building based on understanding the effects of distance and explosive payload on building damage. Figure 5.8 might serve as an approximate guide. We need to know the bollard performance specification relative to the maximum impact energy of a ramming vehicle. This impact energy is derived using combinations of vehicle weight and a scenario-dependent speed of approach.

The kinetic energy of an object is energy acquired through motion. As mentioned in Chapter 2, the formula for kinetic energy is given by $\frac{1}{2} \times \text{mass} \times \text{velocity}^2$, often written compactly as $\frac{1}{2} mv^2$. Again we see scaling featured in assessing and mitigating the vulnerability component of security risk. We notice that kinetic energy scales linearly with mass and nonlinearly with velocity: double the object mass and the kinetic energy is doubled. However, doubling the velocity causes the kinetic energy to quadruple. I feel this effect profoundly when I even slightly increase my pace while running on a treadmill.

There are many examples of moving objects that demonstrate the importance of velocity relative to mass. Defensive backs in football understand this concept quite well. Tackling a big but slow man is a chore, but it is less difficult than tackling a somewhat smaller man running at a much higher speed. High velocity rifle rounds are often more penetrating than slower moving bullets of a larger caliber.

The velocity achieved by an accelerating vehicle is

$$v = v_o + at, \tag{6.6}$$

where v_o is the vehicle initial velocity, v is vehicle velocity, a is acceleration, and t is time. In this analysis we assume the initial velocity is zero. In other words, an approaching vehicle is presumed to charge the building from a dead stop.

The total distance traveled by a uniformly accelerating vehicle is easy to determine from the expression for velocity. This distance x is given by

$$x = \frac{1}{2} at^2 + v_o t. \tag{6.7}$$

For those who know calculus, integrating the expression for acceleration with respect to time yields the expression for velocity, v, and integrating this expression for velocity yields the distance, x. Specifications for vehicle acceleration have been used to plug into the expression for $x = \frac{1}{2} at^2$, assuming the initial velocity is zero.[15]

To assess vulnerability and establish a mitigation strategy for this threat, it is important to know the maximum potential run-up distance by a vehicle. If this run-up distance, x, is known we can solve for t in the expression for the distance. Once the time for an accelerating vehicle to reach a building is known, its velocity on impact can be determined since we have an expression relating velocity, acceleration, and time.

The kinetic energy of a moving object is given by

$$KE = \frac{1}{2}mv^2, \tag{6.8}$$

where m = vehicle mass and v = vehicle velocity.

The vehicle mass and velocity on impact determines the kinetic energy. The kinetic energy for various combinations of mass and velocity that would produce 5.8 MJ (MJ = 10^6 joules) of energy is plotted in Figure 6.7. 5.8 MJ has been chosen because it is an advertised performance rating for one well-known brand of high-performance bollard (Delta Scientific).

We see from Figure 6.7 that a 5000-pound vehicle needs to be traveling a breathtaking 160 miles per hour to achieve a kinetic energy of 5.8 MJ. However, a 70,000-pound tractor trailer must reach a speed of only 43 miles per hour to achieve the same energy. Note that vehicles that weigh more typically accelerate more slowly than lighter vehicles, which affects the required run-up distance. In that vein, to ensure that the bollards satisfy the mitigation requirement as dictated by a given scenario, a mitigation strategy must be designed to ensure a specific vehicle cannot develop enough kinetic energy to exceed the specifications of the

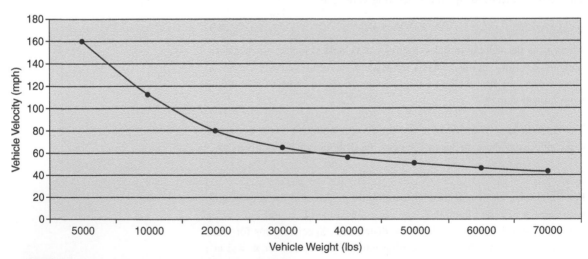

■ **FIGURE 6.7** Vehicle mass-velocity combinations to achieve 5.8 MJ kinetic energy.

proposed bollard. This translates to calculating the available run-up distance, which entails knowing the vehicle's acceleration capabilities vis-à-vis the bollard performance specification.

I have used published vehicle acceleration data for a Ford F-150 pick-up truck[16] to determine the required run-up distance to achieve 5.8 MJ, the performance limit of the bollard referenced previously. The calculation is shown in the following box.

SECURITY METRIC: CALCULATION OF THE REQUIRED RUN-UP DISTANCE FOR A FORD F-150 TO ACHIEVE 5 MJ OF KINETIC ENERGY

In this calculation I determine the required run-up distance to achieve 5 MJ of kinetic energy for a popular American pick-up truck, the Ford F-150. Published data indicate the vehicle can go from 0–60 mph in 8.96 seconds. This works out to a rated acceleration of 3.0 m/s^2 for the vehicle (I assume uniform acceleration). A Ford F-150 weighs about 5500 lb or 2.48 × 10^3 kg.

Since $1/2mv^2$ — kinetic energy = 5.8 MJ, then the required vehicle velocity $v = \sqrt{((2E)/(m))} = \sqrt{((2 \times 5.8) \times 10^6 \text{ kg-m}^2/\text{s}^2))/(2.48 \times 10^3 \text{ kg}))} = 68$ m/s = 153 mph

We know $v = at$, so $t = v/a = 68/3.0 = 23$ sec.

Since we are assuming the vehicle begins from a dead stop (i.e., $v_0 = 0$), the run-up distance, $x = \frac{1}{2}at^2 = \frac{1}{2}(3.0) \times (23)^2 = 794$ m.

Figure 6.8 depicts the run-up distance for various vehicle weights and a 5.8 MJ kinetic energy limit using the method described previously and the acceleration data derived from Reference 15.

If the available run-up distance to a facility is not sufficient for a vehicle to achieve a kinetic energy equal to the energy rating of a particular bollard, then the chosen bollards are up to the job assuming they are installed correctly. Even if the bollards are deemed adequate, consideration should

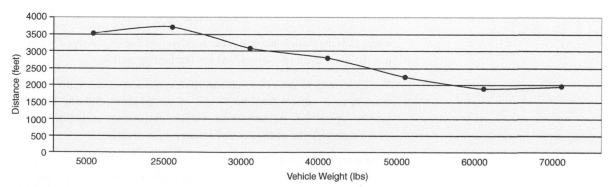

■ **FIGURE 6.8** Run-up distance required to achieve 5.8 MJ kinetic energy.

be given to providing barriers in the direct path of the run-up route (in Britain these are referred to as chicanes). This forces vehicles to follow a circular route and slow their speed or risk tipping over and/or sliding off the road before reaching the target.

Another possible defensive configuration to mitigate the risk of vehicle-borne explosives is the use of a double layer of retractable bollards. In addition to providing significant defensive enhancement to vehicle penetration, a double layer of bollards minimizes the risk of vehicles using piggy-backing to gain unauthorized physical access to a high risk area.

Turning to enter a facility affects a vehicle's maximum speed on approach. A simple relation facilitates the calculation of the maximum turning velocity such that the vehicle does not slip laterally. We assume the road surface is flat and the distribution of vehicle weight is concentrated at its center of mass. This is an idealized situation. It simplifies the problem as it excludes the possibility of torque applied about the center of mass that could cause the vehicle to tip over as it attempts to turn. As is usually the case, this calculation represents an approximation and provides a ballpark estimate of vulnerability to achieve a physically realistic mitigation strategy.

When turning a vehicle, the two opposing forces are friction and centrifugal force. The competition between these forces determines if the vehicle can negotiate a turn and stay within a particular turning radius without slipping. Friction due to the surface area of the tires in contact with the road surface is required to offset the tendency to move laterally as a result of centrifugal acceleration.

We are interested in knowing the maximum velocity for which a vehicle can still maintain its turning radius without slipping. It is clear that the bigger the turning radius r, the greater the velocity that will satisfy this condition. Expressed as an equation, the condition for non-slipping is written as

$$mg\mu = mv^2/r, \tag{6.9}$$

where m is the vehicle mass, g is the acceleration due to gravity or 9.8 m/s^2, μ is the coefficient of friction, and r is the turning radius.

The first thing to notice is that the vehicle mass does not enter into the solution as it cancels from both sides of the equation. Solving for v we get $v = (g\mu r)^{1/2}$. If we assume the vehicle in question uses quality tires, then we specify the coefficient of friction, μ, to be 0.9.

If the entrance to a facility requires a turning radius of 10 m from an access road to enter a facility protected by bollards then the maximum velocity it can maintain without slipping out of its turning radius is $v_{max} = (9.8 \text{ m/s}^2 \times 0.9 \times 10 \text{ m})^{1/2} = 9.4$ m/s, or 21 miles per hour. Increase the turning radius by a factor of 3 to 30 m increases the allowable speed to 36 miles per hour since v_{max} scales as the square root of the turning radius.

Therefore, imposing a small turning radius on vehicles entering a facility from an access road limits the maximum velocity it can achieve and still maintain its course. This in turn limits the maximum kinetic energy developed in attempting to breach the bollards. I must point out that in assuming all the vehicular mass is concentrated at its center of mass I have neglected the real-life effect of a torque produced as a result of circular motion. This real-life effect will further restrict the turning radius of a vehicle and be more pronounced for vehicles with a higher center of gravity.

Roads are sometimes banked so that vehicles can more easily negotiate turns. The condition for constraining the vehicle to the curved path without slipping on banked surfaces is given by tangent $(\theta) = v^2/rg$, where θ is the angle of the banked surface, v is velocity, and r is the radius of the curved path.

The aforementioned elementary analyses are important in determining ballpark estimates on required barrier performance for a given scenario. More complicated analyses that factor in installation details should be performed by qualified engineers to determine the effectiveness of risk mitigation via barriers.

6.5 EXPLOSIVE THREATS

The application of blast film is a traditional method for preventing the splintering of glass and mitigating the hazardous effects of windows subjected to explosive accelerations. The application of blast film is a well-known technique and there are many references in the literature as well as commercial products available. Injuries or death sustained by shards of broken glass are often the most devastating effects of explosive incidents. Following detonation, sharp pieces of window and/or decorative glass are propelled at tremendous speeds exposing anyone in the vicinity to severe injury.

Anti-shatter protection in the form of polyester film is sometimes applied to the glass surface with the intent of preventing splintering. Typical values for film thickness range from 175 to 300 μ, depending in part on the glass surface area and level of protection required. In addition, the glass must be appropriately anchored to its frame for the glass-frame system as a whole to maintain its integrity and withstand the effects of the explosive-induced overpressure and impulse.

The British government provides more detailed performance standards, and The Centre for Protection of National Infrastructure in particular is a source of useful information on window treatments to mitigate explosive risks.[16] A summary of their recommendations is listed in Appendix G. Note that specifications vary depending on the anticipated magnitude of the threat, area of the glass pane, and location within the facility of interest.

Test results conducted by a commercial vendor[17] have demonstrated significant performance enhancements for treated versus untreated glass. For example, untreated quarter inch glass subjected to a 38.9 psi-ms impulse experienced extensive damage, whereas with a 4 mil (1 mil = a thousandth of an inch) anchored film and a similar impulse (39.2 psi-ms) the glass proved safe from breakage.

These tests showed a factor of three enhanced tolerance to peak pressures for windows that were both anchored and glazed versus glazed only. Tests on 14 mil film versus 14 mil daylight (i.e., glazed but unanchored) glass showed a pressure tolerance increase from 3.0 pounds per square inch to 9.1 pounds per square inch.

The effects of explosive loads on buildings is a complex subject, and computer models augmented by experts in this area are required to make true estimates of structural damage in the event of this type of attack. There are a number of firms who specialize in this type of consulting and two well-known companies are Arup Consulting and Weidlinger Associates, Inc., although there are a number of other firms who can provide a similar service.

Notwithstanding the involvement of highly skilled engineers who can leverage sophisticated models, the specific threat scenario must be well understood and precisely specified. This involves articulating the most likely method of explosive delivery and explosive payload plus scenario-driven estimates of the distance between the explosive payload and target. Without this type of preliminary analysis it is impossible to develop a meaningful mitigation strategy.

Equally, detecting the presence of explosive material before detonation necessitates a careful characterization of the site-specific and operational requirements. There continues to be a reliance on specially trained canines for the detection of explosives in commercial settings. There is no doubt that canines can and do detect explosives. But the relevant question is whether they are appropriate to address threats in commercial scenarios of interest.

Surprisingly, this can be an emotive issue. The world of physical security consists of sub-specialists whose passions are ignited by their favorite techniques. Sometimes the effectiveness of these methods is supported by data and other times less so. In addition, canines are cute and their effect on humans is palpable. I recall mentioning to a friend that I was actively pushing to discontinue the wide scale use of explosive detection canines in favor of a combination of methods. It is possible that I would have only gotten a cooler reception if I had confessed to being a mass murderer.

But does this friendly but expensive resource provide value for money in a commercial environment? Again, one must ask the fundamental question regarding the *specific* threat scenarios of interest. Canines have played an extremely valuable role in law enforcement and military applications. But I doubt a dog with even the most sensitive nose would do anything to stop or even deter a person intent on ramming the facility with a vehicle and detonating an explosive payload.

How about using dogs to investigate a suspicious package? Here a dog could add value. But a dog's nose typically becomes "tired." The animal has performance limits beyond which it cannot continuously operate effectively without taking a break. One must question the likelihood that the dog will be available at the required time unless multiple dogs are used. This can be an expensive proposition as special handlers are also required.

Does the use of explosive detection canines act as a deterrent to explosive attacks? Perhaps, but the contention is difficult to confirm. This is one of the catch-all arguments in security that is impossible to prove or disprove and should be viewed with skepticism without justification. If it is indeed true then one might consider deploying dozens of disciplined but untrained canines and raise the defensive profile without the attendant costs. But beware that this strategy could have profound clean-up costs.

There are alternatives or at least adjunct methods in the form of portable and very sensitive explosive detection devices that might be effective depending on the scenario. At the very least, the use of technology could complement the use of canines and possibly facilitate a reduction in the number of required dogs and handlers. Although not as cuddly as a dog, these devices do not get tired and are capable of detecting a range of explosive types. One operational limitation is that surface contact (i.e., swabbing the surface) is necessary to detect non-volatile (i.e., low vapor pressure) explosives, although the device is capable of detecting vapors when they appear in sufficient concentration.

Although the initial expense of a unit may be relatively high, one should do the calculation to compare how quickly a machine would pay for itself when compared to the recurring costs of canines and handlers over time.

■ **FIGURE 6.9** Sabre 4000 explosive detector.

One such device is the Sabre 4000 manufactured by Smiths Detection, a company headquartered in the UK. The company's Web site indicates the device can detect both particulates and vapor, weighs 7 pounds, has a detection time of 10 seconds, yields a complete analysis in 20 seconds, and is capable of detecting the explosives RDX, PETN, TNT, Semtex, TATP, NG, ammonium nitrate, "and others" using Ion Mobility Spectrometry. I have seen a demonstration of this device, and it is compact and relatively easy to use. Figure 6.9 shows a picture of the device taken from the Smiths Detection Web site at http://www.smithsdetection.com/eng/SABRE_4000.php.

Assuming explosives are a threat of concern, I recommend that this device as well as competitors be tested in simulated scenarios before purchasing. It is also important to learn about training, maintenance, and rates of false positives/negatives. A list of customers who have operational experience with this technology would also be invaluable.

6.6 RADIOLOGICAL THREATS

Radiological, biological, and chemical threats or "nontraditional" attacks, as they are sometimes euphemistically known, are thankfully a historical rarity. However, some have occurred and given the potentially lethal consequences of an incident, a prudent mitigation strategy is one where the vulnerability component of risk is understood and an appropriate response protocol is in place. Also, in the event of such an attack there will be a heavy reliance on the response by local, state, and/or federal government representatives and this must be taken into account.

With respect to radiological attacks or radiological dispersion devices, there are many textbooks and Web sites that provide extensive material on the spectrum of potential radiologic agents. I noted in Section 5.5.2 that the deleterious effects of the radiation would likely be minimal, and the explosion represents the most life-threatening feature of an RDD. Nevertheless, it is useful to understand what mitigation procedures would be effective in the event of a radioactive release.

There are two methods to mitigate the effects of a radioactive emitting substance: increase the distance between a radioactive source and absorber and/or shield the source. Figure 5.18 showed the time required for a dose equivalent of 35 rem for ^{137}Cs, ^{192}Ir, and ^{60}Co as a function of the distance from the source. These times are based on the published RHM value. We observed in Chapter 5 that at a distance of 50 m, multiple days of exposure time would be required before clinical radiation effects would be observable for these three radioisotopes.

I also speculated that these same radioisotopes might be considered potential candidates for radiological dispersion devices because of their relative availability, toxicity, and use in half-lives. For the more curious reader, 1000 Ci of these substances weighs 0.9, 11.4, and 2.2 g, respectively (there are 454 grams in a pound). The principal radioactive emissions from these materials are gamma rays in the 0.5–1.5 MeV range.

If we continue to assume little chunks of exploded material will act as point source radiators of energy and neglect the attenuating effect of air, the intensity of radiation diminishes with distance at a $1/4\pi r^2$ rate where r is the distance from the radioactive source. We are assuming the chunks of radioactive material are small compared to the distance from the source. Therefore, the radioactive flux as a function of distance from the source scales accordingly.

In some scenarios it would be reasonable to assume that shielding could be an option for reducing radiation intensity. The highly energetic form of electromagnetic energy emitted from radioisotopes is very penetrating, and typical metal enclosures used to shield radio frequency energy do little to block gamma rays.

The three mechanisms that determine the attenuation of X-rays and gamma rays in matter are the photoelectric effect, pair production, and Compton scattering. Photoelectric absorption occurs when the incoming photon of gamma radiation is absorbed by an electron, and assuming the incoming energy is sufficient, results in the ejection of the electron from its orbit. The theory of the photoelectric effect was first explained by Albert Einstein in 1905 for which he was awarded the Nobel Prize in Physics.

Compton scattering occurs when the incoming photon of radiation is absorbed by an electron and then re-radiated at a different energy by the affected electron. In this process, standard physical principles of conservation of energy and momentum determine the distribution of energies of the scattered photon and recoiling electron.

Finally, when the incoming radiation is sufficient (i.e., twice the rest energy of the electron or 2×0.511 MeV $= 1.02$ MeV), another absorptive process takes place. In this case, the strong electric field produced by the incoming radiation close to the atomic nucleus causes the disappearance of the incoming photon of gamma radiation and the simultaneous creation of an electron and a positron. This process is called pair-production because of this electron–positron "pair" creation.

These three physical effects all contribute to the overall absorption process for X-rays and gamma rays, and the relative magnitude of each are a complex function of the absorbing material atomic number and the energy of the irradiating X-ray/gamma ray photons.

With respect to shielding, lead is not a typical building material so it would be difficult to find a convenient lead shield should the need arise. However, concrete structures are usually in abundant supply and although they are not as effective as lead, they do provide appreciable shielding in their own right.

Figure 6.10 shows the percent attenuation of a 1-MeV energy gamma ray as a function of concrete thickness (in inches). We see from this graph that even a few inches of material has a substantial effect on shielding from gamma radiation (http://www.hanford.gov/alara/PDF/analysis.pdf). The effect of shielding of gamma radiation by lead is calculated in the box titled Security Metric: Lead Shielding of Gamma and X-ray Energy and is intended to illustrate the key concepts. This calculation can be applied to any material using the appropriate values for the mass attenuation coefficient.

With respect to the detection of radiological threats, portable commercial products are available that identify the presence and specific energies of radiological decay products, as well as indicate exposure based on a measurement of intensity.*

Such readings are relatively unambiguous since background levels for gamma radiation and/or neutrons would presumably be significantly less than those encountered during a radioactive release. However, false positives have been known to be triggered by individuals who have undergone certain radiological treatments.

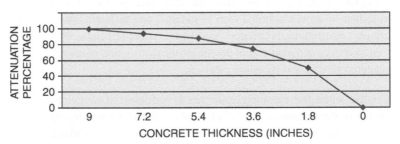

■ **FIGURE 6.10** 1 MeV gamma radiation attenuation through concrete.

---

*Sensor Technology Engineering, Inc., 5553 Hollister Avenue, No. 1, Santa Barbara, CA, produces such a device.

SECURITY METRIC: LEAD SHIELDING OF GAMMA AND X-RAY ENERGY

The change in intensity of X-ray or gamma radiation passing through matter can be characterized by the simple first order differential equation as follows:

$$dI/dr = -(\mu/\rho)I \qquad (6.10)$$

So the change in intensity, I, with distance, r, is reduced by a rate proportional to the mass attenuation coefficient, μ/ρ.

We know from the discussions in Chapter 5, Section 5.5.2 that the solution is a decaying exponential. In fact, we introduced this exact expression in the discussion on radiological dispersion devices in Section 5.5.2.

Specifically, the intensity as a function of distance traveled through the material is given by

$$I = I_0 e^{-(\mu/\rho)\rho r}. \qquad (6.11)$$

I_0 is the original value of radiation intensity, e is the exponential whose value is approximately 2.72, ρ is the material density, μ is the linear attenuation coefficient, μ/ρ is the mass attenuation coefficient at a photon energy in the material, and r is the material thickness traversed by the photon. This expression tells us that the intensity is exponentially attenuated as the energy makes its way through the material.

The published value of μ/ρ can be used to calculate the attenuation caused by the combination of absorption and scattering. This quantity has units of cm^2/gm therefore multiplication by the density of the material is required in solving for intensity or distance.

Lead is a common material used for radiation shielding and this calculation shows why. Ask your dentist about this the next time you get an X-ray and a heavy blanket is placed on top of you before taking pictures of your teeth. It is cheap, dense ($11.3 \ gm/cm^3$) and absorbs X-ray energy in the 1 MeV range mostly due to the photoelectric effect.

We want to calculate the thickness of lead material required to reduce the original intensity of radiation by one-half its initial value. The published value of the mass attenuation coefficient for X-ray energy propagating through lead is $0.06 \ cm^2/gm$ at 1 MeV.

We therefore wish to solve the above relation for the thickness, r, as follows:

$$I/I_0 = \frac{1}{2} = e^{-(\mu/\rho)\rho r}. \qquad (6.12)$$

To "undo" the exponentiation of the exponential function e (an unfortunate coincidence of terminology), we must take the natural logarithm of each side of the equation. This yields the following expression:

$$\ln(1/2) = \ln(e^{-(\mu/\rho)\rho r}) = -(\mu/\rho)\rho r. \qquad (6.13)$$

(Continued)

$$\ln(1/2) = -0.69 = -(\mu/\rho)\rho r$$
$$= -(0.06 \text{ cm}^2/\text{gm}) \times (11.3 \text{ gm/cm}^3) \times (r \text{ cm}) \qquad (6.14)$$

Solving this expression for r yields a figure of about 1 cm. Therefore, a 1 MeV
X-ray beam propagating through lead will lose half its intensity in 1 cm.
Since the attenuation process is logarithmic an additional centimeter of lead
shielding "buys" you ¼ or (½)² reduction in intensity, two more centimeters
yields a 1/8 or (½)³ reduction, etc.

6.7 BIOLOGICAL THREATS
6.7.1 Particulate filtering

Particulate filters are an important building mitigation feature to address
the risk associated with biological threats and also to maintain the general
health of the building population. Particulate filters use fibers to remove
particulates from the airstream. These fibers use three capture mechanisms:
impaction, interception, and diffusion. Interception works as a result of
particle size relative to the fiber that blocks its path. Impaction is where
the particle departs from the airstream by virtue of its momentum and hits
the fiber. Diffusion occurs when the particle trajectory oscillates about
the airstream exhibiting random motion (i.e., Brownian motion). The
larger the particle, the greater the contributions of impaction and interception. Diffusion is actually a more efficient capture mechanism for
smaller particles.[18]

An unambiguous relationship exists between the quality or fineness of particulate filters used in HVAC systems and the vulnerability to building infiltra-
tion by pathogens based on their size. Filters are installed in-line with HVAC
systems to trap incoming airborne particulates before they can enter a
building's ventilation system. It is worth mentioning here that these filters
could also help mitigate the vulnerability component of risk for RDDs by
trapping radioactive dust and reducing the possibility of inhalation or skin
contact, which in addition to the explosion, is the principal concern for such
threats.

HVAC filters are rated for different diameter particle sizes, and their
effectiveness against the spectrum of likely pathogens varies based on
the match-up between filter quality and particulate size. This can be
gleaned from Table 6.2[19] (previously shown in Chapter 5) indicating particle size versus pathogen type in conjunction with the information

Table 6.2 Diameter
of Pathogens

Diameter (μ)	Pathogen Type
< 0.3	Viruses
0.3–1.0	Bacteria, dust, Legionella
3.0–10	Dust, molds, spores
>10	Light pollen, dust mites

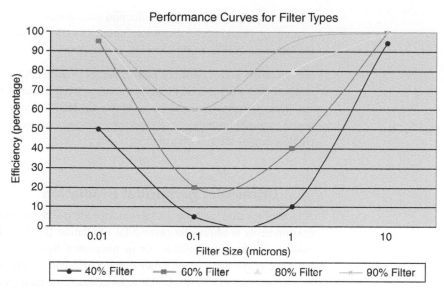

Performance Curves for Filter Types

■ **FIGURE 6.11** Performance curves for particulate filter types.

provided in Figure 6.11,[20] which specifies filter performance as a function of filter size for various filter types.

Notice the dramatic dip in filter effectiveness in the 0.1 to 0.3 μ particle diameter range. This corresponds to the virus-size particle regime so that one would expect that some fraction of an aerosol containing viruses will penetrate these filters and circulate throughout the building courtesy of the HVAC system. The vulnerability of the building population to infection via inhalation versus surface contact and the associated risk mitigation strategy is in part related to the length of time the virus remains in droplet form (see Section 5.5.3) relative to the location of the building occupants. Also, infection rates depend on whether the viruses are released within the building (e.g., via sneezing or by an adversary within the building) or originate externally.

The virus should be more prone to inhalation while still in droplet form since they are carried along in the airstream and supported by the buoyant force of air. Following droplet evaporation, the virus would likely be deposited on surfaces in the form of a salt. The risk of infection in that case results mostly from surface contact, although filters could still help to reduce their dissemination. However, small particles of material would also be subject to the force exerted by the internal air circulation and other physical disturbances. A building mitigation strategy must consider the relative risk for each mode of infection (i.e., surface contamination and inhalation). Section 6.7.5 discusses the relative effectiveness of mitigation for Influenza A viruses for day-long timescales.

For so-called high efficiency particulate air (HEPA) filters, the particle removal rates are extremely high in the 0.1 to 0.3 μ diameter particle range and peak performance is at 0.3 μ. However, HEPA filters can be costly due to required increases in fan and duct size to compensate for restricted air flow across the filter barrier.

6.7.2 Ultraviolet germicidal irradiation

A technology that has been used to enhance the quality of internal building environments is ultraviolet germicidal irradiation (UVGI). This involves subjecting incoming building air to ultraviolet light (C Band or 253.7 nm wavelength) for an appropriate duration and with a minimum intensity. Recent research has shown UVGI to be effective on pathogens of concern in biological attacks.[21,22]

The use of UVGI as a mitigation method can complement particulate filtration by ensuring the spectrum of potential biological agents are either captured or destroyed before they can enter a space. The hope is to obviate the need for high efficiency filtering methods such as HEPA, which are typically necessary to eliminate sub-micron size particles. Even 90–95% filters are reduced in efficiency to 60% and below for particles in the 0.07- to 0.3-μ diameter range (e.g., smallpox, tuberculosis).[23]

The potential utility of UVGI is actually twofold: to provide a healthy internal building environment and also to complement the protective capabilities of HVAC particulate filters in the event of a biological attack. Appropriate intensities of UVGI for a sustained exposure time in conjunction with filter efficiency (i.e., removal rate) have been shown to produce significant reduction in the concentration of harmful pathogens.[24]

Particulate filtration in HVAC air intakes addresses biological and other contaminant threats to varying degrees depending on the filter size relative to particulate diameter. However, these filters have limited effects on chemical releases except for certain aerosols such as tear gases and low volatility nerve agents (e.g., VX).[25]

Figure 6.12 in conjunction with Table 6.3[26] illustrates the effects of particulate filtration for pathogens considered to be potential candidates for biological weapons and also least affected by low efficiency rating filtration.

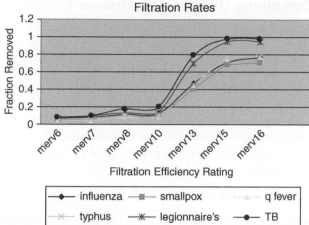

Pathogen	Mean Size (μ)
Influenza	0.098
Smallpox	0.22
Q fever	0.283
Typhus	0.283
Legionnaire's disease	0.520
Tuberculosis	0.637

Table 6.3 Pathogen Mean Sizes

■ **FIGURE 6.12** Particulate filtration rates.

Table 6.4 Removal Rates for Sub-Micron Particles as a Function of Filter Rating (0.3 to 1 μ Particle Size)

MERV 13	<75%
MERV 14	75–85%
MERV 15	85–95%
MERV 16	>95%
MERV 17	>99.97%
MERV 18 and above	>99.99%

The Minimum Efficiency Reporting Value (MERV) designation is used to characterize particulate filter performance. The efficiency of particulate filters for sub-micron particles as specified in the ASHRAE Standard 52.2 is listed in Table 6.4. Filters rated at MERV 17 and above are equivalent to HEPA filters in performance. Although MERV 13 filtration efficiency can be rated at >90%, this rating applies only to pathogens larger than 1 μ in size.

The mean size of some candidate biological weapons is of sub-micron proportions. MERV 13 filters have less than 75% efficiency for pathogens of this size as noted in Table 6.4. Pathogens and toxins near or above 1 μ in diameter (e.g., *R. rickettsii*, also known as Rocky Mountain spotted fever and *Bacillus anthracis*, anthrax and botulism) are removed at rates equal to or greater than 90% using MERV 13 filters. Can UVGI provide enhanced removal/kill rates for sub-micron pathogens and thereby obviate the need for HEPA filters?

6.7.3 **Combining UVGI and particulate filtering**

The pathogen kill rate from UVGI is given by[27]

$$KR = 1 - e^{-kIt} \qquad (6.15)$$

Here k is an experimentally determined rate constant ($cm^2/\mu W$-sec) that differs for each pathogen, I represents the average UV intensity ($\mu W/cm^2$), and t is the pathogen transit/exposure time (seconds). The overall kill rate resulting from the combination of filtering and UVGI cannot simply be added since the second method in the sequence of elimination methods operates only on the surviving population of pathogens. The overall kill rate is therefore given by

$$KR = 1 - (1 - KR_1)(1 - KR_2) \qquad (6.16)$$

KR_1 and KR_2 are the kill/removal rates for filtration or UVGI, respectively, depending on the sequence of implementation. This expression was used to determine the set of curves shown in Figure 6.13[28] that denotes the total kill rates using a combined UVGI and particulate filtration system.

These curves indicate improvement in sub-micron pathogen reduction when UV and filtration methods are combined relative to the exclusive use of non-HEPA particulate filtration. Specifically, in excess of 84%

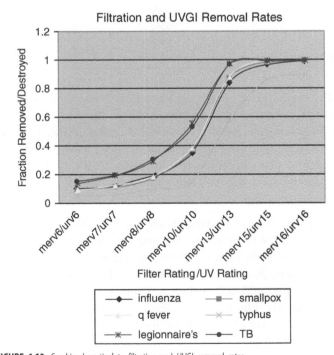

■ **FIGURE 6.13** Combined particulate filtration and UVGI removal rates.

removal rates can be achieved for a MERV13/URV13 rated system and above for all listed pathogens except typhus. The UV intensity ratings (URV) are shown in Table 6.5.[29]

The pathogen removal rates shown in Figure 6.14 assume particle velocities of 2.54 m/sec within the HVAC intake chamber. They also assume a 0.5-sec exposure time. The 2.54-m/sec figure is posited as an average HVAC air velocity value, and is a good approximation if one assumes laminar flow (i.e., uniform distribution of air velocities, low Reynolds number) in the HVAC chamber.

If it is assumed that the average air velocity is 2.54 m/sec and a 0.5-sec exposure time, which are considered standard system parameters,[30] this imposes a path length requirement for UV exposure of at least 1.27 m since path length = 2.54 m/sec × 0.5 sec. UVGI installations could be influenced by this parameter since the required path length might be constrained by the available real estate.

However, in real life the air movement within HVAC air intakes is turbulent and causes considerable air mixing. The extent of air mixing on the effectiveness of UVGI is not well understood, although some particles will likely experience increased UV exposure as a result. The so-called Reynolds number,* which indicates the level of turbulent flow, has been estimated to be 150,000 in commercial HVAC systems,[31] which represents significant turbulence.

It has been theoretically demonstrated that air turbulence in commercial HVAC systems could play a small role in the effectiveness of UVGI.[32] Because of the statistical nature of particles carried along in turbulent flows, random particulate motion could actually increase the requirement for UV exposure for some pathogens, which translates to increasing either the UV fluency or the path length.

Developing an effective building defense against nontraditional threats is difficult, and this discussion is only meant to introduce the reader to some of the salient issues associated with this problem. Qualified building engineers should be consulted in conjunction with security experts to recommend an appropriate defense against particulate-based building contamination. As always, it is incumbent upon the security professional to make appropriate decisions on mitigation based on an understanding of the components of risk and mitigation options.

Table 6.5 URV Ratings (µW)	
URV6	75
URV7	100
URV8	150
URV10	500
URV13	2000
URV15	4000
URV16	5000

---

*The Reynolds number is a dimensionless number used in fluid dynamics that is a measure of the ratio of inertial forces to viscous forces. A high Reynolds number implies significant turbulence.

6.7.4 **More risk mitigation for biological threats**

Biological weapons can also be introduced into a building environment via letters or packages. In 2001 – anthrax spores were sent to government personnel via the US mail resulting in five deaths. There are pathogen detection schemes available, and these are typically based on either immunoassay or DNA amplification techniques (i.e., polymerase chain reaction; PCR).

Each of these techniques has its advantages and disadvantages, but neither is instantaneous and they currently require minutes of processing time. Immunoassays are known to have a higher false positive rate than PCR, and DNA amplification is limited in the number of pathogens it can detect, although such systems are advertised by vendors to have a 2×10^{-6} false positive rate.

With respect to filtration, a high performance electrostatic filter has been developed by Strionair (www.strionair.com). In this device, the particles in the air and the filter fibers become charged and are mutually attracted in strong electric fields. The company claims near-HEPA performance for sub-micron particles in the electrical enhancement mode without the associated pressure drops that accompany HEPA filters. The Strionair device also purports to both capture and kill pathogens, which are especially nice features if your job is to clean and/or change building filters. According to their Web site, independent laboratory tests have been conducted that confirm advertised results.

Figure 6.14 illustrates the advertised filter results from Strionair.

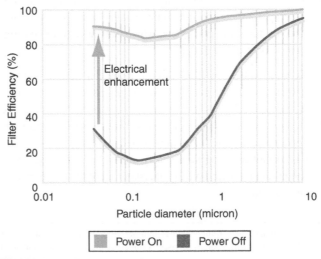

■ **FIGURE 6.14** Strionair filter advertised performance.

Biological agent detection schemes are available that sample the environment and use particle sizing to alert occupants to the presence of threats. These have a reputation for producing false positives due to the natural presence of harmless organisms in the air. It has been suggested that another method of detection might be to detect the *rate* of increase of airborne particles since a precipitous rise in particle number might signify the presence of a credible threat and reduce the risk of false positive alerts, but I do not know of any such detector at this time.

Finally, dual energy or Compton backscatter X-ray machines should be considered for detecting low density material that might contain biological agents concealed in mail or packages. These machines exploit the fact that X-ray energy is absorbed and/or scattered preferentially depending on the material density and the X-ray energy. These two phenomena were discussed in Chapter 5 in connection with radiological dispersion devices. The photoelectric effect and Compton scattering are the causes of absorption and scattering of X-rays, respectively, for energies below 1 MeV which is where typical commercial X-ray units operate.

I conducted a very informal experiment using a dual energy device that showed that a relatively small amount of low density powder placed in an envelope could be detected among a stack of similar envelopes containing paper. Based on this admittedly nonrigorous result, I am optimistic that this might have potential as a relatively easy screening method to detect biological materials concealed in envelopes.

6.7.5 **Relative effectiveness of influenza mitigation**

Although building mitigation methods for airborne pathogens such as ventilation are well known, their *relative* effectiveness would be a useful bit of information in developing a mitigation strategy. Models of airborne infection in confined spaces have generally been based on the notion of a unit of infection called a "quantum" introduced by Wells in 1955.[33] A quantum is used to express the response of susceptible individuals to inhaling infectious droplet nuclei. Wells postulated that not all inhaled droplet nuclei result in infection and defined a quantum of infection as the number of infectious droplet nuclei required to infect $1-1/e$ susceptible people.

The Wells-Riley model[33,34], which is based on the notion of a quantum, has been used to predict the number of new infection cases, N_c, over a period of time t (seconds), in an indoor environment ventilated at constant rate Q (m^3/s), and is given by

$$N_c = S(1 - e^{-Iqpt/Q}) \qquad (6.17)$$

Here S represents the number of susceptible people in the space, I is the number of infectious people, p (m^3/sec) is the pulmonary ventilation rate of susceptible individuals, and q represents the quantum unit of infection defined above. This equation is typically used to predict average infection risk in various scenarios, although numerical simulations are required in the case of small populations such as those confined to an indoor space.

The Wells-Riley equation coupled with actual exposure data for tuberculosis was used to determine the probability of infection as a function of room ventilation. These results are plotted for Intensive Care Unit (ICU) and office building environments in Figure 6.15.[35,36]

Although plotted together, the two curves cannot be compared directly because the actual ventilation rate per occupant and room air exchanges resulting from the ventilation in each venue are different. However, security professionals are often interested in the results for office buildings. According to this model, increasing total ventilation for office buildings using realistic air flows would have marginal effects on infection rates. For example, increasing ventilation from the existing 15 cubic feet per meter (cfm) outdoor air per occupant by 33% would protect only a few more of the 27 workers infected in the actual exposure according to the model. Even doubling the ventilation rate to 30 cfm, a highly unusual value for an office building environment, would have protected only about half of those infected.

However, it would be useful to understand the effect of ventilation in conjunction with other forms of risk mitigation. Metrics that enable an understanding of the *relative* effectiveness of mitigation measures would provide a quantitative foundation for a mitigation strategy. In that vein,

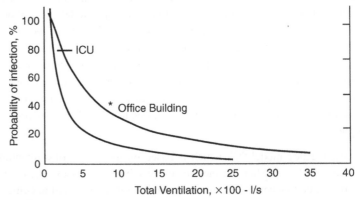

■ **FIGURE 6.15** Probability of tuberculosis infection versus ventilation rate (indoors) using the Wells-Riley model. *From Nadell FA, Keegan J, Cheney SA, Etkind SC, American Review of Respiratory Disease, 1991; 144:302–306.*

an attempt was made to perform such an analysis of office environments for the H5N1 virus or bird flu over day-long time frames, where the quantum concept was abandoned and where the following factors were taken into account: (1) surface and airborne viruses, (2) air turbulence, and (3) the effect of surface deposition of airborne viruses as well as direct surface contact by infected individuals.[37] This work has natural extensions to contagious biological weapons, although the specific values of parameters vary according to the pathogen of concern.

With respect to Influenza A (note H5N1 is a type of Influenza A virus), contracting this illness depends on the susceptibility and degree of infection of an individual, the infection profile of the population entering and leaving the room, and the virus dynamics within the room. The adopted approach was dependent on (1) determining the time course of the virus and the susceptibility of an individual, (2) assuming an infection profile of the population entering and leaving the room consistent with the local population, and (3) understanding the dynamics of the virus in the room relative to anti-virus measures (exclusive of anti-viral medications).

To understand the effects of the virus on an individual, six species of cells were analyzed to create equations describing their evolution in time. Scaling laws were developed for variations in (a) incubation time, (b) the maximum number of infected cells, and (c) the duration of the infection by approximating a previous model using six ordinary differential equations. Dependence of a, b, and c on the initial viral load was shown to be slow: linear with the logarithm of virus number and not varying at all over many orders of magnitude until a critical viral load is reached.[38]

Next, and using simple models of air turbulence in a room as well as the results previously cited, an equation for the infection density in a room, $i(t) \sim i_0 e^{Rt}$, was developed. In this expression R = areal infection density growth rate and t = time. Although this expression looks deceptively simple, R actually turns out to be a function of a number of growth and damping terms with airborne and surface virus parameters related to the scaling laws for the effects of a virus on an individual. The threshold condition for infection spreading in a room is that the growth term exceeds the damping term.

The analysis yielded an expression for the infection density, i(t), and suggested the following results regarding the relative effectiveness of mitigation for Influenza A virus suppression in a facility:

- As the room occupation level increases, the building ventilation air velocity required to suppress room infections increases.

- The building air ventilation velocity required to suppress infection increases markedly if vented air is partially recirculated in a poorly filtered condition.
- Enhanced virulence increases the required building ventilation air velocity.
- Required building ventilation air velocity variation with room occupancy is linear up to 1 m^{-2} (in the absence of face masks or gloves); low room occupancy is also an effective prophylactic measure.
- For face mask filter efficiencies above 0.5, the required vent air velocity is low; the use of proper masks and gloves is a highly effective prophylactic measure.
- Low vent air velocity in conjunction with high room occupancy greatly increases the importance of the air filtration efficiency and the air recirculation fraction.

The following graphics (Figures 6.16–6.19) summarize the quantitative results of this analysis. The relative values of specific mitigation measures are compared directly to understand which combinations of conditions lead to non-growth of infections within a room over day-long time scales.

Specifically, Figure 6.16 shows the required threshold for virus growth in a room as a function of ventilation velocity and room population density (no mask and gloves). The bottom, middle, and top curves correspond to 0.1, 0.5, and 0.9 system parameters, respectively, where this system parameter is equal to the product of f_R (the fraction of room air recirculated through the input vents) and f_{VS} (the surviving fraction of viruses after passing through any filter and/or UV system in the ventilation mechanism). For velocities above the curve the virus does not propagate. Note that the lower the value of the system parameter, the greater the effectiveness of filtering and air recirculation. Therefore, the lower the

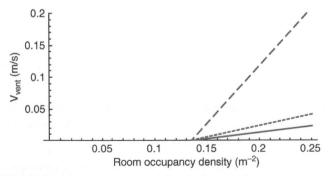

■ **FIGURE 6.16** Threshold for virus infection growth (no masks or gloves).

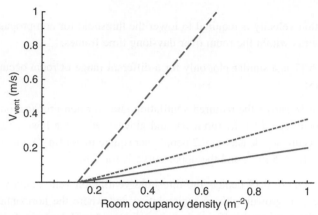

■ **FIGURE 6.17** Threshold for virus infection growth (increased room occupancy density and no mask or gloves).

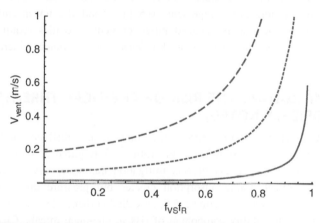

■ **FIGURE 6.18** Threshold for virus infection growth (no mask and gloves).

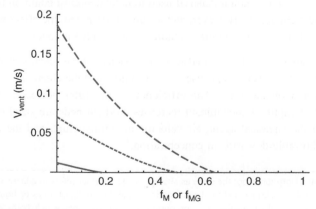

■ **FIGURE 6.19** Threshold for virus infection growth (with masks or masks and gloves).

ventilation velocity is required to lower the threshold for non-propagation of the virus within the room over day-long time frames.

Figure 6.17 is a similar plot only for a different range of room occupancy densities:

Figure 6.18 shows the required ventilation rates for non-propagation plotted directly against $f_r f_{vs}$ (no mask and gloves) where the lowest curve is for an occupancy density of 0.2 people per square meter (m^{-2}), the middle curve is for 0.5 m^{-2}, and the upper curve is for 1 m^{-2}.

Finally, Figure 6.19 depicts the threshold ventilation vent velocity (m/s) for infection growth with no recirculation and where the horizontal axis shows the face mask efficiency f_M or the combined face mask, glove, and hygiene efficiency, f_{MG}. It has been assumed that $f_M = f_{MG}$. The lowest curve is for an occupancy density of 0.2 people per square meter (m^{-2}), the middle curve represents 0.5 m^{-2}, and the upper curve is 1 m^{-2}. Note the greatly reduced range of vent velocities required to suppress infection growth as the mask or mask–glove efficiency increases.

6.8 **MITIGATING THE RISK OF CHEMICAL THREATS (BRIEFLY NOTED)**

Mitigating chemical or vapor-borne threats are difficult problems for facilities because significant adsorption* is required at building air intakes. The effectiveness of this mitigation can also vary depending on the chemical. Large adsorption creates big pressure drops across filters. As discussed in Section 5.5.5, so-called sorbent filters are used to reduce the vulnerability component of risk to chemical threats. Charcoal (e.g., copper-silver-zinc-molybdenum-trithylenediamine; ASZM-TEDA) in particular is a common material used to defend against common battlefield chemical agents. However, and as previously noted, the effectiveness of the sorbent depends on the contaminant and ambient conditions.

Sorbents are rated in terms of adsorption capacity for the specific chemical of interest. The adsorption capacity is the amount of the chemical that can be captured by the filter. The efficiency of the filter decreases as the amount of captured contaminant increases. Sorbent beds are sized on the basis of the chemical agent, air velocity and temperature, and the maximum allowable downstream concentration.[39]

---

*Note that adsorption is not the same as absorption. The former refers to adherence by a gas or liquid to a surface and the latter implies to the dissolving of a gas or liquid in a medium. The level of adsorption is proportional to surface area which helps explain the enhanced load imposed by sorbent filters on building HVAC systems.

There are a number of factors that affect sorbent filter performance and the decision to install such filtration should be done in consultation with a bona fide expert. Maintaining such filters is non-trivial and requires specific knowledge, since other chemicals and humidity can affect the bed capacity. It may be more practical to adopt a strategy of containment, where high risk areas such as the lobby or mail sorting room are addressed separately via dedicated mitigation.

Deploying separate air handling systems and establishing negative pressure relative to the rest of the building should be considered for areas deemed at high risk. In the event of a chemical or other toxic gaseous release, fresh air can be brought in from the outside to dilute the effects of the contaminant without polluting other areas of the building. If the area is at negative pressure with respect to the other portions of the building but positively pressurized relative to outdoors, the risk of contaminating other areas is reduced.

A relatively low number of infectious biological particulates can have the desired deleterious effect since individuals become ill from a relatively small infecting dose and then spread it to others. Biological attackers hope to disperse their products far and wide through contagious mechanisms. Therefore, the attack strategy is to avoid redundant contacts of the pathogen with the same individual to infect the maximum number of individuals using a finite amount of material. With respect to mitigation, dispersal of the material using dilution could be contraindicated since more people might actually be impacted.

In contrast, a chemical attack requires a minimum concentration to perform the desired effect, although the bad news is this minimum concentration may be quite low. Therefore, dilution with air to a level below the minimum concentration might be effective in reducing the overall impact of a chemical attack. Once released into the atmosphere, chemicals must overcome the effects of evaporation and/or dispersal to ensure an adequate concentration over a specified area.

The potential for a chemical attack inside a commercial facility is difficult to assess. Unless demonstrated to the contrary with credible intelligence, it is probably not a threat worth expending huge resources relative to other threats. If such an event occurred, quickly segregating the affected area in combination with increasing the flux of uncontaminated air to dilute the effect of the chemical might be a practical strategy. This may represent the best one could do absent a large and possibly disproportionate expenditure of resources.

6.9 **GUIDELINES FOR REDUCING THE VULNERABILITY TO NON-TRADITIONAL THREATS IN COMMERCIAL FACILITIES**

As noted throughout this book it is important to assess the individual components of risk to develop practical security strategies. In the end, businesses must be financially competitive so it will do no good to expend resources that jeopardize the financial well-being of a company to reduce vulnerability to relatively unlikely threats.

Therefore, and as always, the security professional must use his judgment to inform management on a *reasonable* course of action based on a realistic and commercially driven assessment of the vulnerability and likelihood components of risk. It is in that spirit that the following list of building features and general security procedures is presented to help reduce vulnerability to biological, chemical, and/or radiological threats.

1. Place external air intakes as high above ground level as possible. This is perhaps the most important of all mitigation with respect to chemical, biological, or radiological threats.
2. Filter incoming air using particulate matter filters using the highest rated (i.e., MERV rating) filter considered feasible and practical for a particular system and in consideration of likely threats.
3. Provide separate air handling systems for all publicly accessed areas including lobbies, reception areas, mail handling, or package delivery facilities and provide for physical isolation from the general building population. Design these areas so they are at negative pressure relative to the rest of the building, but at positive pressure relative to the outdoors. Design common return-air systems or return pathways (e.g., ceiling plenum) so that they are not shared with other areas of the building.
4. Provide individual air handling systems with a dedicated on/off capability.
5. Use ducted air returns.
6. Secure return air vents from unauthorized physical access.
7. Cover externally accessible external air vents. Protect these from unauthorized physical access.
8. Design buildings for maximum air tightness.
9. Restrict access to building air handling units and lock/monitor physical access to entrances.
10. Mail and packages should be screened away from business areas and ideally be located off-site.
11. Properly authenticate and authorize physical access by all personnel.

12. Develop and train personnel on emergency response procedures and provide an area for "sheltering in place" if possible.

6.10 **COMMERCIAL TECHNICAL SURVEILLANCE COUNTERMEASURES**

The loss of sensitive information is an important concern for commercial firms. Retaining a competitive advantage and avoiding embarrassing publicity are also quite important. Loss of intellectual property and/or time-sensitive information could be devastating to the bottom line as well as to a firm's reputation. Industrial espionage represents one highly publicized threat related to information loss. Commercial technical surveillance countermeasures (TSCM) are a popular method employed by companies to reduce the vulnerability component of risk with respect to this threat.

Commercial TSCM employs methods to detect the presence of technologies that are designed to covertly gather signals that contain information of interest. One such example that made its way into the media was the discovery of technology deployed aboard Air France jets presumably to detect conversations between executives of international corporations.*

Some investigative firms offer TSCM services to clients and commercial TSCM practitioners are often advertised as former government (a variety of governments) employees with relevant experience. The use of commercial TSCM is prevalent throughout the private sector yet its capabilities and inherent limitations are sometimes not well understood by corporate security professionals. Candidly, TSCM vendors often do not help this situation.

Companies sometimes utilize commercial TSCM as the principal form of mitigation to reduce the vulnerability to the threat of corporate espionage. The methods employed are almost viewed as magic by some executives who pay handsomely for this service. This near-mystical view is encouraged by vendors and is enhanced by a reliance on sophisticated electronic widgets coupled with a lack of transparency with respect to their effectiveness.

Although commercial TSCM is touted as a panacea for industrial espionage, it is also sometimes viewed as a perfunctory box to tick in the spectrum of available mitigation. After all, no one wants to be second guessed if a bug had actually been found and commercial TSCM had not been utilized. In my view, the following questions need to be addressed to have confidence in commercial TSCM: (1) is there statistical confidence that

---

*"Air France Denies Spying on Travelers," *New York Times*, September 14, 1991.

commercial TSCM as it is currently practiced is effective, (2) are there alternative (and possibly cheaper) inspection methods that might be equally effective or at least complement and/or enhance its effectiveness, and (3) how can TSCM be successfully integrated into a broader program of information management?

Anyone who liked the 1970s ultra-paranoid movie *The Conversation* cannot help but be enamored with commercial TSCM, yours truly included. Shrouded in mystery, it appeals to those of us who are susceptible to the allure of a shadowy world and nifty electronic gizmos. On the surface it is a highly efficient technique since it appears to reduce the risk of a priority threat in one fell swoop. Unfortunately, reality is more complicated.

It has been historically difficult to accurately gauge the effectiveness of commercial TSCM. Reliable statistics on its success are not generally available and vendors are notoriously fuzzy on this point. It must be noted that this is not a regulated industry, so anyone can set up shop and start debugging everything in sight. This is not to say that the current commercial TSCM methods cannot be successful on any given day. But even a blind (but not necessarily psychotic) squirrel finds a nut occasionally. Consistency, rigor, flexibility, and intellectual honesty are the characteristics we seek here.

So what are some of the issues with commercial TSCM as it is currently practiced? For one, inspections of companies tend to follow patterns. Whether these are performed at night, on weekends, every 6 months, etc., any hint of regularity constitutes a pattern that can be exploited. Vendors are not necessarily at fault here since corporate executives typically want their cake and eat it too. They demand a secure environment but will not tolerate being disturbed or inconvenienced so that only certain inspection times are allowed.

An adversary with physical access to sensitive locations and who knows that commercial TSCM inspections will be conducted according to a schedule has a definite advantage. This includes knowing when inspections will *not* be conducted. Once an inspection pattern becomes known, a wily adversary could merely remove the device or shut it down when the risk of discovery is deemed unacceptably high.

It is also important to recognize that the instant after a space has been inspected it can no longer be considered pristine if physical access to that space is not restricted to trusted individuals. This argues for using commercial TSCM in a focused manner to protect specific rooms at required times. Finally, TSCM vendors have a habit of ignoring fundamental scientific and operational constraints in favor of cookie cutter procedures and the reflexive use of high-tech inspection devices.

A lack of inspection predictability might be more concerning to a would-be industrial spy than the specter of sophisticated counter surveillance technology. The positive news is that conducting TSCM inspections with the requisite attention to scenario-driven realities and in the absence of a discernable pattern has the potential to yield dividends.

Let's examine some practical details associated with devices that broadcast audible information in the form of radio frequency signals and are commercially available. The Spy Shop is one well-known purveyor of such devices. Reliably transmitting low-level, radio frequency signals to remote locations may appear easy in the movies, but the situation is actually more complex.

A discussion of how radio frequency signals are attenuated and reflected by materials in and around buildings, especially in dense urban environments, was included in Chapter 5. Based on these data, unless an adversary is willing to broadcast a big signal and risk detection, transmitting low-amplitude signals over appreciable distances could be a problem. An adversary is subject to the same physical laws that govern signal propagation as the rest of us. It is worth thinking about the proximity of nearest neighbor buildings or the closest space to your office that is not under your physical control when assessing the vulnerability to attacks using radio frequency transmitters.

Also, and thanks to the often lax behavior of employees in handling sensitive documents, it may not be necessary to resort to radio frequency transmitters to obtain information from sensitive areas. If an adversary has regular physical access to areas where information is stored, created, or discarded then this opens up a world of unpleasant opportunities and could obviate the need for the use of technology to do his dirty work. This lesson is sometimes lost on even sophisticated security types: *allowing only properly authorized and authenticated individuals to access sensitive areas represents the single most important countermeasure to the threat of information loss and other types of losses as well.*

Even from a strictly technical perspective, broadcasting a radio frequency signal seems an unnecessary risk for the modern industrial spy thanks to advances in commercial electronics in just the last few years. Consider the following: widely available memory technology exists in the form of 4 GB flash memory cards (e.g., see www.sandisk.com). These memory devices are the size of a matchbook, weigh about 0.5 ounces, and retail for approximately $1000 each. Assuming the collection of telephone-quality speech is the goal, this implies a minimum 6 kHz sampling rate (i.e., 3 kHz signal bandwidth and Nyquist sampling). Using 8-bit quantization of the audio signal, this works out to 96×10^3 bits/sec.

Therefore utilizing a single, 4 GB memory card as part of a recording device would yield 83 hours or ~10 business days (assuming an 8-hour business day) of speech. Compression might extend this time by a factor of 5 or more (digital cameras use 5:1 compression ratios), and multiple memory cards would seem to be an option. So there appears to be little reason to transmit a potentially compromising and finicky electromagnetic signal assuming an adversary has even modest physical access to areas of interest. Using a commercial recording device and periodically dumping the contents of memory would accomplish the intended goal with relatively low risk.

In fact, it has been suggested that a recording device consisting of a microphone, audio and control electronics, 128 Mb memory, and lithium cell power source could some day be constructed that was less than one cubic *millimeter* in volume. This device would provide a full eight hours of sound and would be no bigger than a large grain of salt. Although it is not realistic to expect that such a device would be commercially available anytime soon, it serves to illustrate the point that through advances in electronics, broadcasting a radio signal should become an increasingly unnecessary risk and technical burden to an adversary with physical access to sensitive areas.

Providing physical access to a location where sensitive information is discussed, created, stored, or discarded to a less than trustworthy individual is a recipe for problems on a number of levels. This is also an obvious risk factor for the theft of physical items as well as for information, as I found out the hard way in my office. It is also important to keep in mind that the risk associated with the threat of information loss is not a one or zero proposition. Common sense dictates that options for device concealments grow in proportion to an adversary's unsupervised time on site. If physical access to sensitive areas is limited then so are the opportunities for bad things to happen.

Although they are not sexy from a technical perspective, simple visual searches might add value if such searches are conducted frequently, unpredictably, and the searcher is instructed on anomalous features. Looking for out-of-place changes provides clues that someone may be up to no good. Of course careful documentation is required to be able to recognize such changes, which could be subtle. Careful documentation of the results of TSCM inspections should be a *sine qua non* of any commercial TSCM program.

Unscheduled and frequent inspections by appropriately trained members of the on-site security force might also act as a deterrent to any potential industrial spy. This form of mitigation might reduce an adversary's confidence that a device can be retrieved or shut down before it is discovered in addition to increasing the chances of catching an adversary red-handed.

Moreover, readily available and inexpensive technology might be exploited to detect anomalous room features in countries where traditional commercial TSCM inspection equipment is prohibited (e.g., Russia) or because of budgetary constraints. For example, simple electronic stud detectors that can be purchased at a hardware store might be useful in locating hidden and out-of-place objects. Readers who are DIY-oriented are probably already familiar with these devices. They emit a radio frequency signal and detect the reflected signal to highlight differences in the capacitance of dissimilar materials due to associated changes in the dielectric constant. This is why the metal stud evokes a different response than wood or wallboard. No special training would be required except to indoctrinate the user on what constitutes an anomalous response and where to focus attention. In addition, X-rays of incongruous or new objects could instantly reveal the presence of extraneous and unwanted surprises.

The use of poorly shielded cell phones or AM/FM radios to detect the presence of active digital devices might also be applicable. Recall the annoying presence of a Blackberry when it is brought in close proximity to some telephones or radios. Portable magnetometers are a readily available and inexpensive option to detect the presence of anomalous metal objects. Relatively inexpensive spectrum analyzers or other radio frequency detection devices could be effective in detecting signals emanating from strange places. If the signal power mysteriously decreases by $1/r^2$ as you move away from that strange place, it could be a hint that there may be a problem.

Signal amplitudes close to a transmitting device significantly reduce the requirement for detection sensitivity as I demonstrate in the box titled Security Metric: The Effect of Physical Proximity in Detecting a Radio Frequency Signal. Although such devices are not necessarily broadband in their frequency response, it is not clear that a broadband capability is required for many physically realistic scenarios. In that vein, commercial TSCM vendors sometimes tout their ability to detect radiating devices over a wide range of the electromagnetic spectrum: "From DC to

Daylight" is the mantra I sometimes hear from commercial TSCM vendors in connection with their detection capabilities.

However, in view of the frequency dependence of intra- and inter-building electromagnetic propagation where high frequencies are preferentially attenuated, it would make little sense to look for signal transmissions at frequencies that would be significantly attenuated by intervening structures and/or over significant distances. Calculations and information to support this contention are well represented in this book.

On a related topic, the limits associated with sending information via digital transmission channels are well understood and characterized by the Shannon-Hartley Law. This says that the encoded communication rate or digital channel capacity is given by $C = W \log_2 (1 + S/N)$, where S/N is the linear expression (i.e., not in dB) for the S/N ratio, $\log_2$ is the logarithm in base 2, and W is the channel bandwidth. The units of C are in bits/second.

Therefore for any communication channel, the rate of information transfer is dictated by the bandwidth and S/N of that channel. So it is less likely to find devices that must transmit high bandwidth signals via channels that can only accommodate low frequency transmissions.

The take-away from the aforementioned discussion is that planning a countermeasures strategy must factor in the realities of the physical world. Ultimately it is physics in combination with physical access that dictates what is both possible and practical for any potential adversary and no technology has been invented that can fool Mother Nature.

This is not to say that standard commercial TSCM inspection equipment cannot be effective when judiciously and intelligently deployed. For example, nonlinear junction detectors (NLJT) as advertised on the Internet emit a radio frequency signal and detect the presence of so-called nonlinear electronic components such as diodes. These are potentially quite useful precisely because they (again) are able to detect anomalous features, e.g., the presence of out-of-place electronics. Also, a hidden device does not have to be powered or radiating to be located using this technique.

But in general there appears to be a near-reflexive response by commercial TSCM vendors to apply sophisticated and expensive methods without due consideration of scenario-specific constraints and the realities of the physical world. There are definitely times when more sophisticated inspection methods are required. But equally, value might be added by complementing traditional TSCM visits with frequent and unscheduled inspections using inexpensive equipment capable of detecting anomalous features. No one can guarantee that this "poor man's inspection regimen" will be as effective as one that uses traditional inspection equipment. But the converse cannot be proven either. Unfortunately there is a paucity of statistical data on commercial TSCM results and this represents a significant issue in my view.

Furthermore, I believe there is value to be gained in experimenting with inexpensive methods for use in commercial TSCM inspections. In addition, these can potentially be used as part of a program of self-inspections by willing clients. The effectiveness of such methods will likely improve with regular and methodical (but not predictable) use, and of course, assiduous record keeping. It is worth experimenting with these techniques by both vendor and client alike, and some progressive vendors may already be doing this. A more focused inspection using traditional techniques by legitimate experts could be used to confirm an anomalous finding before tearing the CEO's office apart.

Commercial TSCM vendors who cloak their methods in secrecy, describe their instruments as black boxes, are reluctant to share even sanitized data, and are generally not forthcoming about their methods should be avoided. An experienced commercial TSCM provider should agree to train your security staff to use certain tools and methods based on their aptitude and enthusiasm. On the other hand, a commercial TSCM vendor that shows a willingness to explore the use of alternate methods and encourages client participation as desired should be given a closer look.

The advantage of measuring radio frequency signals close to the transmission source and the implications to inspection technology is demonstrated in the following section.

SECURITY METRIC: THE EFFECT OF PHYSICAL PROXIMITY IN DETECTING A RADIO FREQUENCY SIGNAL

Consider a commercially available, relatively low-power, off-the-shelf, 10 mW microphone and radiofrequency transmitter located somewhere in a room. The commercial TSCM inspection vendor team is armed with a simple radio frequency signal analyzer (assumed to operate in the frequency range of the transmitting device). While moving the analyzer antenna around the room they fortuitously position the antenna approximately one meter from the transmitter. What is the power as detected by the analyzer assuming isotropic radiation and unity antenna gain?

The total energy of the electromagnetic field of the radiating transmitter is the sum of the electric field (E) and magnetic field (B) energy densities:

$$\text{Energy} = \tfrac{1}{2}\varepsilon_0 E^2 + (\tfrac{1}{2})(1/\mu_0)B^2 \tag{6.18}$$

ε_0 is the electric permittivity in free space (8.85×10^{-12} F/m) and μ_0 is the magnetic permeability in free space ($4\pi \times 10^{-7}$ webers/amp-meter).

(Continued)

SECURITY METRIC: THE EFFECT OF PHYSICAL PROXIMITY IN DETECTING A RADIO FREQUENCY SIGNAL—CONT'D

Since $B = E/c$ (c = the speed of light = 3×10^8 m/s) and $c = (1/\varepsilon_0\mu_0)^{1/2}$ then

$$\tfrac{1}{2}(1/\mu_0)B^2 = \tfrac{1}{2}(1/\mu_0 c^2)\varepsilon_0 E^2 = \varepsilon_0\mu_0 E^2/2\mu_0 = \tfrac{1}{2}\varepsilon_0 E^2 \qquad (6.19)$$

The electromagnetic field radiation intensity is denoted as I. Half of the radiated intensity is derived from the electric field and half from the magnetic field. The radiated intensity has units of watts per meter squared (W/m²). At a distance r from the transmitter the radiated intensity, I = radiated power/$4\pi r^2$ which equals the radiated energy times its speed of propagation (i.e., the speed of light = c).

So,

$$\tfrac{1}{2}I = \tfrac{1}{2} P/4\pi r^2 = \tfrac{1}{2} c\varepsilon_0 E^2$$
$$P/4\pi r^2 = I = 0.8 \times 10^{-3} \text{W/m}^2 = \varepsilon_0 E^2 c \qquad (6.20)$$

where $c = 3 \times 10^8$ m/s

Therefore,

$$E = (0.8 \times 10^{-3}/\varepsilon_0 c)^{1/2} = 0.55 \text{ V/m} \qquad (6.21)$$

Note that TV and FM signals are typically in the mV/m or μV/m range so the electric field of the hidden transmitter as measured at one meter is approximately a thousand to a million times greater in magnitude. This should give some comfort to those opting for relatively inexpensive detection gear.

Assuming a 0.5-m unity gain antenna on the detection device yields an induced voltage of 0.3 V. If we assume 50 ohms for the power meter input impedance R, and noting Power = V^2/R, then the signal power from the hidden transmitter as seen by the commercial TSCM inspector is a healthy 1.8 mW.

Compare this with the same calculation at 25 m where an adversary is hanging out with her receive gear, neglecting the attenuating effects of intervening structures. The signal power she measures using the same setup is about 2 μW (i.e., 2×10^{-6} W) or roughly a thousand times less power than the signal received by the commercial TSCM inspection team. The signal is still easily detectable at that distance, but the point is that even relatively insensitive equipment could do the job assuming an appropriate frequency response by the detection device.

Of equal concern to the loss of information via the use of spooky technology is the loss or leakage of information through building design vulnerabilities or by employees who do careless and/or irresponsible things.

Consider a conference room in which sensitive information is discussed and where its structural elements conduct sound to uncontrolled external space and/or to other portions of the building. In Chapter 5 the vulnerability of audible information to structure-borne propagation of acoustic energy was discussed. A complete commercial TSCM inspection regimen should include assessing and reporting any such vulnerability.

The employee who decides to establish a "rogue" LAN for his convenience despite a company policy forbidding such networks could represent a problem. Commercial TSCM could be an extremely effective means of identifying such activities since the frequencies of transmission are well-known and regularly scheduled inspections would not be a problem in this instance.

On a programmatic level, commercial TSCM vendors should not be viewed as the exclusive purveyor of anti-information loss or leakage solutions. Information protection is a multifaceted issue and can just as easily be caused by carelessness as well as theft. It therefore requires a combination of technical methods, training, and awareness via education. The restriction and control of physical access to sensitive areas in addition to managing the creation, handling, and destruction of physical documents represent key elements of a truly effective information protection program.

Finally, the selection of an appropriate commercial TSCM vendor can be difficult in light of a lack of both industry regulation and available performance data. Therefore, questions that might be included in a request for proposal are listed in the following questionnaire. A vendor's response to some or all of these could be used to evaluate its philosophy as well as highlight its relative strengths and weaknesses. In general, a TSCM vendor should be evaluated not only by its technical prowess but also in its overall approach to the more general and complex problem of protecting sensitive and confidential information.

6.10.1 Questionnaire for prospective commercial TSCM vendors

1. How long has your company been performing countermeasures surveys?
2. Characterize the nature of your company's experience in electronics, signals, and electronic surveillance.
3. What type of formal engineering/technical training does your countermeasures personnel possess?
4. Describe your company's general approach to electronic countermeasures as part of a broader program to counter information loss.
5. Have you ever provided training in countermeasures to a client's security staff to augment more formal technical inspections?

6. What equipment do you typically deploy in performing countermeasures surveys?

7. Have you built and/or designed/adapted any of your own equipment? Please describe.

8. Indicate the types of *confirmed* covert devices you have located with this equipment and/or other inspection techniques utilized. Do you keep statistics on devices so deployed? Have you observed any trends in the commercial use of electronic devices to obtain unauthorized information?

9. Do you perform audits or "red-teaming" exercises (i.e., blind testing)? (Note: this is an extremely revealing indicator. I would personally include any vendor on a short list that demonstrated a commitment to rigorous red-teaming of their procedures and then admitted to a less than perfect rate of success.) Are the results of such audits documented?

10. Are your reports standardized or are they customized to meet specific client requirements? May we see a sample of your written product?

11. How do you remain current on technical threats relative to specific industries or countries?

12. What background checks are performed on employees in your company, and how often are these background checks updated?

13. Please indicate a list of references.

Finally, the inherently spooky nature of commercial TSCM makes it both interesting and offbeat. It is in that spirit that I recommend viewing a "Spy versus Spy" cartoon from *Mad Magazine* that captures the lighter side of this form of risk mitigation and can be found at www.bugsweeps.com/info/spytech.html.

6.11 **ELECTROMAGNETIC PULSE WEAPONS**

Concerns about the threat of non-nuclear electromagnetic pulse (EMP) weapons directed against commercial facilities have emerged into the public consciousness. These weapons generate a high amplitude and broad band electromagnetic pulse. Energy so developed is intended to inductively and/or capacitively couple to electronic circuits with potentially disruptive or catastrophic results.

A number of potential mechanisms are available for delivering high intensity electromagnetic energy in scenarios of concern. These include explosively pumped flux compression generators (EPFCG), explosive and propellant driven magneto-hydrodynamic generators, and high power microwave sources.[40] This crude analysis focuses on EPFCG as the weapon of choice. However, the generation of an electromagnetic pulse is common to all EMP devices and this pulse is the source of damage to electronic circuitry independent of its mode of generation.

6.11.1 **The EPFCG threat**

EPFCG devices operate by using a conventional explosive to rapidly compress a magnetic field, transferring energy from the explosive to the magnetic field itself.* These devices are reportedly capable of producing peak power levels of tens of terawatts (10^{12} W). For reference, the peak power level of a lightning stroke can be as much as a terawatt.

The scenario considered here is a vehicle containing an EPFCG device detonated in the vicinity of a data center. Upon detonation, a powerful electromagnetic pulse is generated that consists of a broad spectrum of frequencies. In theory, an appropriately constructed Faraday cage (i.e., an appropriately sealed metal box) could effectively shield circuits from external sources of electromagnetic energy. However, it is important to appreciate that electronic devices require leads for power and information transfer that preclude complete physical isolation from the outside world. Therefore, multiple and complementary methods are required to establish an effective mitigation strategy.

6.11.2 **EMP generated in proximity to unshielded facilities**

In this scenario it is assumed that no shielding exists around servers, routers, and other technical equipment, and that underground cables feeding data centers are not affected by the EMP. Specifically, an electromagnetic pulse generated by an EPFCG directly interacts with the internal circuitry of key electronic components in digital devices. Such circuitry is comprised of Metal Oxide Semiconductor (MOS) devices that are sensitive to high voltage transients.

Power cables and interconnections among these electronic circuits develop potentials in response to a changing magnetic flux. This voltage will cause arcing through the insulation that separates the gate from the drain/source elements of field effect transistors (FET). This results in the breakdown of the device. Voltage levels so induced would be applied to the gates of the FET, etc. Such devices have maximum safe operating voltages listed at about ±20 V. Other electronic components not considered here have even smaller breakdown voltages.

---

*The physical principle exploited in these devices is known as Lenz's law and it is an example of the conservation of energy.

Consider an EPFCG that generates a terawatt of electromagnetic power detonated in the vicinity of a data center. I will perform a rough calculation of the electric field and induced voltage caused by a hypothesized weapon at a representative distance (one kilometer) from the pulse source. The calculation is nearly identical to the one performed in Section 6.10 where I calculated the electric field from a radio frequency transmitter and Equation 6.17 represents the starting point for both calculations.

The energy of the pulse propagates with equal intensity in all directions (i.e., isotropically) unless specifically designed to do otherwise. The energy density of an electromagnetic wave is given by the following expression:

$$\text{Total Energy Density} = (1/2)\varepsilon_o E^2 + (1/(2\mu_0))B^2 \qquad (6.17)$$

ε_o is the electric permittivity of free space, μ_0 is the magnetic permeability of free space, and E and B are the electric and magnetic field amplitudes, respectively.

It can be shown that the energy is divided equally between the electric and magnetic fields. The electromagnetic field intensity, I, caused by the pulse has units of watts per square meter (W/m^2) and is equal to

$$I = \varepsilon_o E^2 c \qquad (6.18)$$

Recall that c is the speed of light.

As noted above, the power of the electromagnetic pulse following the explosion is 10^{12} W. The intensity or power density as a function of distance for a wave propagating isotropically is $P/4\pi r^2$, where r is the distance from the explosive source.

Equating the power density of the propagating pulse (moving at the speed of light) with the intensity of the electromagnetic field yields

$$P/4\pi r^2 = \varepsilon_o E^2 c \qquad (6.19)$$

The electric field at a distance of one kilometer from the source is now easily calculated to provide insight into the effectiveness of physical separation from the vehicle and the facility. Note that it might be difficult to establish even one kilometer of physical standoff for most facilities, and this rough calculation suggests this distance does not offer much protection.

The distance, r, is 10^3 m and plugging this value into the above expression and using $c = 3 \times 10^8$ m/s, $\varepsilon_o = 8.9 \times 10^{-12}$ F/m, the electric field is calculated to be 5.5×10^3 V/m. Therefore, a cable of any significant length interfaced to a rack of electronics would be vulnerable to kilovolt level potentials. This greatly exceeds the breakdown voltage of internal digital components.

Even if this calculation is in error by an order of magnitude based on poorly chosen parameter values and/or erroneous physical assumptions, such voltage levels would exceed electronic component operating specifications.

6.11.3 **EMP generated in proximity to shielded facilities**

Here we assume the target facility or components within the facility are electromagnetically shielded using an appropriate metal enclosure. The effect of shielding is to create eddy currents (in response to magnetic fields . . . Lenz's law again) and currents (in the case of electric fields) that cancel the effect of the electromagnetic radiation. In this scenario, a Faraday cage is presumed to protect electronics so enclosed from the effects of direct EMP radiation.

I leverage the many EMP events of which we are all familiar to help to appreciate the effect of an EPFCG-derived pulse for this scenario. These events are cloud-to-earth lightning strikes. The first stroke of lightning during a thunderstorm can produce peak currents ranging from 1000 to 100,000 amperes with rise times of 1 μs.

Despite the presence of shielding, it would be difficult to defend a facility with sensitive electronics against a stroke of this magnitude at close range. An EPFCG source might be expected to produce currents equal to or exceeding those expected from lightning and be spatially non-localized.

However, buried cables represent the more likely vulnerability as induced voltages on these elements are more common for lightning scenarios. Experiments have been conducted that yield estimates of peak voltages at varying distances from a cable. In one such test, lightning-induced voltages caused by strokes in ground flashes at distances of 5 km (3.1 miles) were measured at both ends of a 448-m long non-energized power distribution line. The maximum induced voltage was 80 V peak to peak.[41] Using simple and now familiar $1/r^2$ scaling arguments would suggest a factor of 25 increase in induced voltage at a distance of only 1 km.

Recall that tens of volts presented to input gates of FET devices exceed specified breakdown voltages. Eighty volts clearly exceeds such levels although surge protection for voltages of this magnitude is commercially available.

The point is that data centers are also susceptible to detonated EMP devices because of power cables feeding typical facilities even if electromagnetic shielding is employed. EMP pulses generated at distances of a few miles from a facility could in theory be defended using surge protection and/or isolation transformers.

It is also important to note that even if it is assumed that electronic devices are contained within shielded enclosures, small openings such as those surrounding wires entering the enclosure pass electromagnetic energy at wavelengths shorter than the size of the opening (think about the fact that one can peer inside a microwave oven while electromagnetic radiation cooks your turkey burger).

Although information on the bandwidth of EPFCG-derived pulses is not readily available, presumably a short duration pulse could produce energy at wavelengths of a few tenths of a millimeter (i.e., microwaves), but this requires confirmation. This is small enough to allow energy to leak into the gap created by wire insulation. The moral of the story is that effective mitigation would mandate that care must be taken to ensure the installation of robust electromagnetic shields.

Standoff distances required to ensure adequate separation between an EMP source (i.e., a vehicle containing an EPFCG device) and a target facility are considerable to mitigate the risk of catastrophic damage to electronic components. This is based on the vulnerability to voltage and current surges generated by direct irradiation and/or signal coupling to underground cables.

A combination of shielding, grounding, and surge protection might provide effective mitigation for some EMP scenarios. As usual, the security professional must make a cost-benefit analysis and examine the trade-off of the potential for such an attack against the expense of risk mitigation. A qualified power systems engineer should be consulted to scope the problem for a specific facility and determine the practicality and cost-effectiveness of proposed solutions.

6.12 SUMMARY

Analyses of simple mitigation strategies were provided in this chapter to address the threat of information loss from unauthorized detection of acoustic and electromagnetic signals. These involve the use of barriers made of various materials and evaluated at a range of signal frequencies. The effectiveness of such strategies is highly frequency- and materials-dependent, and there is significant variation in the estimated effectiveness of the designed mitigation depending on the scenario.

The threat of vehicle-borne explosives is a serious one and there have been many incidents around the world. Vulnerability to such threats is a function of standoff distance as well as the quantity and type of explosive. Barriers and bollards are typically used to enforce a minimum vehicular standoff distance. Information is provided to determine the crash-tested kinetic energy requirement for such barriers based on scenario-specific parameters.

Nontraditional threats considered high impact but low likelihood, e.g., biological, chemical, and radiological attacks, represent an increasing concern to security professionals. Mitigating these threats can be difficult, but existing measures include building filtration for biological attacks and sorbent filters for chemicals. UVGI can be used to supplement particulate filters to mitigate the risk from a broad range of pathogens. Developing appropriate plans for sheltering-in-place with inherent limitations as noted herein, creating and practicing evacuation plans, and coordinating with local authorities are important components of any risk mitigation strategy.

Commercial TSCM is sometimes viewed as a panacea for countering information loss from the threat of industrial espionage. An analysis of the effectiveness of traditional commercial TSCM is difficult without reliable performance data. Simple technical and procedural arguments reveal concerns with commercial TSCM as it is traditionally practiced. It is recommended that inspections focus on frequent but unscheduled searches and complemented by visual inspections. In addition, consideration should be given to experimenting with nontraditional inspection methods as described in this chapter. TSCM should be considered to be one element of a broader program to counter the threat of information loss and leakage.

Finally, EMP weapons represent another variety of nontraditional threat, and an elementary calculation of vulnerability and associated analysis of the required mitigation is provided. Furthermore, this analysis reveals the inherent difficulties in mitigating the vulnerability component of risk associated with this threat.

REFERENCES

1. Hall Donald E. *Basic Acoustics*. Harper and Row; 1987.
2. Ibid.
3. www.engineeringtoolbox.com.
4. http://www.squ1.com/index.php?http://www.squ1.com/sound/absorption.html.
5. Hou P. *Investigation of the Propagation Characteristics of Indoor Radio Channels in GHz Wavebands*. Goettingen: Cuvillier Verlag; 1997.
6. Hashemi H. The Indoor Radio Propagation Channel. *Proceedings of the IEEE*. 1993;81(7).

7. White Donald R. *A Handbook on Electromagnetic Shielding, Materials and Performance*. Fourth Printing; 1986.

8. http://www.chomerics.com/tech/Shielding_methods.htm.

9. Ibid.

10. Ibid.

11. http://www.sss-mag.com/indoor.html.

12. Katz R. Radio Propagation. In: *Computer Science*. Berkeley: University of California; 1996:294–297.

13. Ibid.

14. Ibid.

15. http://www.wpoplin.com/Acceleration_of_Heavy_%20Trucks.pdf.

16. http://www.PickupTrucks.com.

17. http://www.blastgard.com.

18. http://www.asbestosguru-oberta.com/hepa.htm.

19. Department of Health and Human Services, Centers for Disease Control and Prevention, National Institute of Occupational Safety and Health (NIOSH). *Guidance for Filtration and Air-Cleaning Systems to Protect Building Environments from Airborne Chemical, Biological, or Radiological Attacks*, April 2003.

20. Ensor DS, Hanley JT, Sparks LE. *Filter Efficiency*. Washington DC: Healthy Buildings/IAQ; 1991.

21. Brickner P, Vincent R, First M, Nardell E, Murray M, Kaufman W. The Application of Ultraviolet Germicidal Irradiation to Control Transmission of Airborne Disease: Bioterrorism Countermeasure. *Public Health Rep*. 2003;118.

22. Kowalski WJ, Bahnfleth WP. Immune Building Technology and Bioterrorism Defense. *HPAC Engineering*. 2003.

23. Ensor DS, Hanley JT, Sparks LE. *Filter Efficiency*. Washington DC: Healthy Buildings/IAQ; 1991.

24. Kowalski W, Bahnfleth W, Musser A. Modeling Immune Building Systems for Bioterrorism Defense. *Journal of Architectural Engineering*. 2003;9(2).

25. *Guidance for Filtration and Air-Cleaning Systems to Protect Building Environments from Airborne Chemical, Biological or Radiological Attacks*. National Institute for Occupational Safety and Health (NIOSH), Centers for Disease Control; April 2003.

26. Kowalski W, Bahnfleth W, Musser A. Modeling Immune Building Systems for Bioterrorism Defense. *Journal of Architectural Engineering*. 2003;9(2).

27. Ibid.

28. Ibid.

29. Ibid.

30. Kowalski WJ, Bahnfleth WP, Witham DL, Severin BF, Whittam TS. Mathematical modelling of ultraviolet germicidal irradiation for air disinfection. *Quantitative Microbiology*. 2:249–270.

31. Ibid.

32. Chang D, Young C. Effect of Turbulence on Ultraviolet Germicidal Irradiation. *Journal of Architectural Engineering*. 2007.

33. Wells WF. *Airborne Contagion and Hygiene*. Cambridge, MA: Harvard University Press; 1955.

34. Riley EC, Murphy G, Riley RL. Airborne Spread of Measles in a Suburban Elementary School. *Am J Epidemiol*. 1978;107:421–432.

35. Noakes CJ, Sleigh AP. Applying the Wells-Riley Equation to the Risk of Airborne Infection in Hospital Environments: The Importance of Stochastic and Proximity Effects. *Indoor Air*. 2008.

36. Nardell EA, Keegan J, Cheyney SA, Etkind SC. Airborne Infection. Theoretical Limits of Protection Achievable by Building Ventilation. *Am Rev Respir Dis*. 1991;144.

37. Chang D, Young C. The Relative Effectiveness of Anti-Influenza Measures in an Office Environment. 2007 (unpublished).

38. Chang D, Young C. Simple Scaling Laws for Influenza A Rise Time, Duration and Severity. *J Theor Biol*. 2007;246.

39. *Guidance for Filtration and Air-Cleaning Systems to Protect Building Environments from Airborne Chemical, Biological or Radiological Attacks*. National Institute for Occupational Safety and Health (NIOSH), Centers for Disease Control; April 2003.

40. Kopp C. The Electromagnetic Bomb — A Weapon of Electrical Mass Destruction. http://www.globalsecurity.org/military/library/report/1996/apjemp.htm.

41. Schneider K. *Lightning and Surge Protection*. USA: Telebyte.

Epilogue

Hopefully after reading this book the reader appreciates the breadth and logic of security risk management. Although security is a field that is varied in scope, and the problems it addresses are sometimes unique, the methodology used is the same as any other field that involves managing risk.

It is precisely this breadth that mandates participation by individuals from very different professions to solve real problems. Modern security risk management problems require contributions from scientists with specialized knowledge in partnership with security professionals who bring relevant experience and judgment to the table.

However, these seemingly strange bedfellows would benefit from a common framework with which to apply their collective knowledge, skills, and experience. The main goal of this book has been to provide that framework and to introduce some of the important tools to facilitate dialogue and collaboration.

The practice of security is fundamentally about managing the risk associated with threats. Risk is the lens through which security must be viewed irrespective of the viewing angle if it is to have applicability and relevance. Furthermore, an assessment of the risk associated with each threat should be at the heart of every security decision. So understanding the individual components of risk and the factors that influence those components are essential to developing effective risk mitigation strategies. This book has devoted significant attention to conveying a deeper understanding of the likelihood and vulnerability components of risk pursuant to facilitating such strategies.

Understanding the components of risk is a necessary but not sufficient condition for effective security risk management. The modern security professional requires analytic tools with which can apply judgment to a structured but flexible risk mitigation process.

This book offers a methodology for risk mitigation in the form of controls, methods, and performance criteria that is applicable to any security risk management problem. Security standards, audits, risk metrics, and program frameworks are shown to be natural by-products of this methodology. Examples are provided to illustrate the concept and its broad applicability.

Assessing risk in any more than a superficial way requires an appreciation for some common mathematical and scientific concepts. Recurring physical models and processes offer important insights into the likelihood and vulnerability components of risk and facilitate more quantitative approaches to risk. These models were introduced in Part I and explained further in Part II of this book. They are expressed in terms of standard mathematical terms to characterize physical quantities that affect risk.

In summary, the objects of this book are to enable scientists to see the applicability of potentially familiar concepts and to enlighten security professionals with respect to the more technical dimensions of security risk management.

Bringing the worlds of science and security together is a challenge. But I firmly believe the game has been worth the chase. Despite my earlier cynicism about the resilience of marriage, I am convinced that the connection between scientists and security professionals is becoming increasingly important. My hope is that this book contributes to a lasting and fulfilling relationship.

Scientific prefixes

Prefix	Symbol	Value
exa-	E	10^{18}
peta-	P	10^{15}
tera-	T	10^{12}
giga-	G	10^{9}
mega-	M	10^{6}
kilo-	k	10^{3}
hecto-	h	10^{2}
deca-	da	10^{1}
deci-	d	10^{-1}
centi-	c	10^{-2}
milli-	m	10^{-3}
micro-	μ	10^{-6}
nano-	n	10^{-9}
pico-	p	10^{-12}
femto-	f	10^{-15}
atto-	a	10^{-18}

Sound levels and intensities

Source of sound	Intensity level (dB)	Intensity (W m^{-2})	Perception
jet plane at 30 m	140	100	extreme pain
threshold of pain	125	3	pain
pneumatic drill	110	10^{-1}	very loud
siren at 30 m	100	10^{-2}	
loud car horn	90	10^{-3}	loud
door slamming	80	10^{-4}	
busy street traffic	70	10^{-5}	noisy
normal conversation	60	10^{-6}	moderate
quiet radio	40	10^{-8}	quiet
quiet room	20	10^{-10}	very quiet
rustle of leaves	10	10^{-11}	
threshold of hearing	0	10^{-12}	

■ **FIGURE B.1** Approximate sound levels and intensities within human hearing range.

The speed of sound in common materials

Material	Speed of Sound (m/s)
Steel	6100
Aluminum	4877
Brick	4176
Hardwood	3962
Glass	3962
Copper	3901
Brass	3475
Concrete	3231
Water	1433
Lead	1158
Cork	366
Air	343
Rubber	150

Closed circuit television (CCTV) performance criteria and technical specifications

PERFORMANCE CRITERIA

(a) Facial Identification: identify individual(s) from the displayed/
recorded image

or

(b) Situational Awareness: discern actions/events within a defined area of
coverage from the displayed/recorded image

OPERATIONAL MODES

(a) Fixed position with adjustable focus
(b) Pan-Tilt-Zoom (PTZ)

IMAGE DATA AND TRANSMISSION REQUIREMENTS

(a) Internet Protocol (IP)
(b) Network Connectivity

CAMERA/SYSTEM MANAGEMENT

(a) Network video recorder (IP)
(b) Digital video recorder

IMAGE RESOLUTION

(a) 4 Common Intermediate Format (CIF)
(b) Adequate to achieve the stated performance criterion

RECORD FRAME RATE

(a) Minimum of five frames-per-second
(b) Adequate to achieve the stated performance criterion

IMAGE STORAGE

Thirty-one (31) days subject to local statutory/regulatory requirements or otherwise specified by the security director based on stated investigative requirement

AMBIENT LIGHTING

(a) Minimum three foot-candles ($\sim$30 lux) over the field of view
(b) Adequate to achieve the stated performance criterion

POWER AND RESILIENCE

(a) Local mains with UPS
(b) Battery back-up with capacity dictated by requirements

FIELD OF VIEW

(a) Between 40 and 62 degrees (36–60 mm focal length)
(b) Special lenses with focal length and associated depth-of-field based on stated performance criterion

INFORMATION SECURITY RESTRICTIONS

All IP-based components must be installed within company-controlled space.

Physical access authorization system performance criteria

HIGH-LEVEL SYSTEM ARCHITECTURE

(a) Globally distributed
(b) Centrally and regionally monitored
(c) Common software application management

PHYSICAL ACCESS AUTHORIZATION

(a) Company ID card authorization token
(b) Approved ID card technology only
(c) Authorization for physical access granted via link to company employee database
(d) Company ID-compatible card reader linked to locally programmable physical access authorization panel
(e) Recommended maximum of 8 to 16 card readers per panel
(f) Visible confirmation of physical access authorization at card reader (e.g., green or red LED)
(g) Potential enhancement to two-factor authentication (e.g., ID card plus PIN pad)
(h) ID card readers installed for "in" and "out" authorization/registration in sole tenant facility lobby turnstiles

PHYSICAL ACCESS AUTHORIZATION CONDITIONS AND SIGNALING

(a) Signal confirming authorized physical access linked to magnetic door-locking mechanism or lobby turnstiles
(b) Alarm and alert signal types include "door forced" and "door held open"; specific alarm/alert parameters to be determined by the security director
(c) Unique physical access authorization code for each restricted area

PHYSICAL ACCESS AUTHORIZATION INFORMATION TRANSMISSION

(a) Hard wire connection from card reader to panel
(b) Network connectivity from panel to physical access authorization server
(c) Physical access authorization servers and other network-linked security equipment managed by appropriate company technical entity

PHYSICAL ACCESS AUTHORIZATION HISTORY AND REPORTING

(a) Record of all authorized physical access histories and signal conditions
(b) Five-year secure storage of physical access histories
(c) Customized reporting of physical access histories

PHYSICAL ACCESS AUTHORIZATION EQUIPMENT SECURITY

Physical access authorization panels security in compliance with the Physical Security Standard for Office Facilities

Exterior barrier performance criteria and technical specifications

1. Barriers should circumscribe the base of the facility if protecting the entire facility.
2. Barrier emplacement should be at the maximum allowable distance from the facility or the minimum required distance to minimize blast effects according to a proper risk assessment.
3. Maximum spacing between barrier elements dictated by kinetic energy requirements (vehicles should impact a minimum of two barriers).
4. Retractable barriers should be used to facilitate legitimate vehicle physical access at specific facility locations (e.g., loading bays).
5. Retractable barriers should be remotely operable (i.e., operated from a location away from the barrier).
6. Kinetic energy specification based on maximum estimated vehicle run-up velocity per US Department of State (DoS) Specification (K-value):
   a) Less than 1 m penetration for a 15,000-lb (6804 kg) vehicle impacting the barrier at 30 miles per hour (48 km/hr) (K4).
   b) Less than 1 m penetration for a 15,000-lb vehicle impacting the barrier at 40 miles per hour (64 km/hr) (K8).
   c) Less than 1 m penetration for a 15,000-lb vehicle impacting the barrier at 50 miles per hour (80 km/hr) (K12).
7. Minimum barrier height is 3 feet (900 mm). Recommended barrier height is 3.9 feet (1200 mm).

Window anti-blast methods technical specifications*

I. Polyester Anti-Shatter Film Applied to Glass Panes
    A. 175 µ thickness for standard window panes.
    B. 300 µ thickness for panes over 6 square meters, over 8 mm thick, or ground floor windows.
    C. 100 µ thickness if used in conjunction with blast net curtains
    D. Apply film to the glass to extreme edges for new windows or areas being re-glazed before fixing frames.
    E. Maximum gap of 3–5 mm between film and frame if fitting to windows *in situ*.
    F. For double glazed windows consisting of two separate and independently operating frames, apply film to both panes,
    G. For double glazed windows where the inner pane cannot be opened independently or a "sealed unit" is fitted, apply film to inner pane only.

II. Blast Net Curtains
    A. Use only in conjunction with anti-shatter film.
    B. 90 or 100 denier polyester terylene material.
    C. Curtains should be twice the width and 1.5 times the length of the glass pane.
    D. Bottom hem must incorporate flexible weights at the rate of 400 g/m.
    E. Excess length should be folded concertina-wise and placed in shallow troughs at window sill level.
    F. 50–100 mm distance from the glass pane if possible.
    G. Flame resistant to a minimum standard determined by local requirements.

---

*Reproduced from the British Security Service Web Site.

III. Blast-Resistant Glass (laminated)

1. Laminated glass is preferable to tempered glass.
2. 7.55 mm minimum thickness including a 1.5 mm minimum thickness layer of polyvinylbutryal (PVB).
3. Pane should be fixed in a frame to withstand bending effects of 7 kN/square meter over the complete area of the glazing and frame.
4. Rebates fixed to the frame and of the frame to the building structure should be designed to withstand line loads of 20 kN/m along the perimeter (based on requirements for two square meter glass pane area).
5. Loading requirements should be scaled up by 50% to match the increased blast resistance of smaller window panes (i.e., about one square meter).
6. Line loads for fixings/attachments may be factored down by 25% to 15 kN/m along the perimeter for windows of about four square meters overall area. However, the 7 kN/square meter requirement in #3 above should remain in force.
7. Glass panes with an edge dimension of 1 m or more should be provided with a frame having a glazing rebate of at least 35 mm giving a bearing of 30 mm. (Note: greater protection may be provided by setting the pane in double-sided adhesive security glazing tape or ideally bonded in sealant.)
8. If robust frames and deep rebates cannot be provided, a level of protection equivalent to anti-shatter film on plain glass with net curtains can be achieved using thinner laminated glass, e.g., 6.8 mm thick.
9. In double glazing, the preferred standard is 7.5 mm laminated glass inner pane with a 6-mm tempered glass outer pane in a robust frame with deep rebates.
10. For laminated glass in standard frames, the laminated inner pane may be reduced to 6.8 mm (with 0.76 mm PVB) and a 4-mm tempered outer layer if panes are less than two square meters in area. The strength of fixings to the frame should be designed to resist no less than 5 kN/m. Fixings should be installed at a minimum separation distance of 350 mm centers.

Qualitative interpretation of Rw values

25 = Conversation level speech is easily understood
30 = Loud speech is easily heard and understood
35 = Loud speech can be heard but not understood
42 = Loud speech is heard as a murmur
45 = Straining is required to hear loud speech
48 = Loud speech is almost inaudible
50 = Loud speech is inaudible

Index

Note: Page numbers followed by *b* indicate boxes, *f* indicate figures and *t* indicate tables.

Printed and bound by CPI Group (UK) Ltd, Croydon, CR0 4YY

03/10/2024

01040343-0008